2 · 4)

(47

LIFE BEYOND

FROM PRISON TO MARS

LIFE BEYOND

FROM PRISON TO MARS

Editor: Charles Cockell

The British Interplanetary Society
London

Life Beyond - From Prison to Mars
Editor: Charles Cockell

Published by:
The British Interplanetary Society
    Arthur C Clarke House
        27-29 South Lambeth Road,
            London, SW8 1SZ, UK
www.bis-space.com

This publication is financed by The British Interplanetary Society which is a charity, registered in the UK (No: 250556). Any residual profit from the sales of this book will be used to support the Society's work to promote and advance space education and outreach.

First edition: July 2018

ISBN- 9781983289088

Produced by Amazon.com Inc. Kindle Direct Publishing

Cover design Alex Storer
www.thelightdream.net

LIFE BEYOND - FROM PRISON TO MARS

Contents

INTRODUCTION:
LIFE BEYOND – FROM PRISON TO MARS

Charles Cockell

UK Centre for Astrobiology, School of Physics and Astronomy,
University of Edinburgh, Edinburgh, EH9 3JZ

One of the most exciting frontiers for humanity is the exploration and settlement of space. Over the last few years, private companies have begun to explore space alongside growing government involvement. Now a large collection of private and public organisations has begun launching robots and advancing plans to send a greater number of people into space.

This growing activity offers tremendous opportunities to release people's creative capacities. With this in mind, in 2016 the UK Centre for Astrobiology, in collaboration with the Scottish Prison Service (SPS) developed a program called *Life Beyond* to actively engage the prison population in the design of settlements for Mars. The program began in 2016 with a pilot scheme across four prisons in Scotland (HMP Edinburgh, Glenochil, Lowmoss and Shotts) to refine the course content and establish the best way to implement the course. The pilot scheme resulted in the construction of a four-week course. In the first week, the *Life Beyond* course starts with participants considering what challenges there are to living on Mars. In the second week, they transition into designing their own Mars station, incorporating the challenges identified in the first week, such as oxygen and food production, into the station design. A two or three-week break at this point in the course gives time for participants to work on their designs. In week three, they consider what the occupants at the station would do, such as planning expeditions from the station or conducting science. In the final week, they consider how a society would be run on Mars. What challenges in governance, civic responsibilities and management confront a group of people living in a station on the Red Planet? They can use their own experiences in prison to think about some of these issues, stimulating discussion on interdependence of societies, democracy and active citizenship.

The program has had spectacular results. Two courses run at HMP Glenochil and Edinburgh resulted in the Mars station designs presented in this publication. One of the most satisfying opportunities that the course offers is for a group of participants to work as a team in designing their base, but within that team each individual can pursue their own interests and capacities. Some of the participants have focused on engineering, others on art and others on human factors. In HMP Glenochil, one participant with an interest in music engaged two friends and produced a 'Martian Blues' song, an impressive synthesis of blues music with science fiction lyrics and Zulu chants. One day, when explorers are sitting on Mars listening to blues music, they should remember that the first blues song written for future Mars explorers was composed at HMP Glenochil, Scotland.

The point of the *Life Beyond* project is very simple – from behind the boundaries of a prison, you can direct humanity to the stars. The work undertaken by the participants in this course has been impressive by any standards, particularly since what is presented here was the result of a four-week effort. Many of the participants had no prior expertise in space exploration. We hope that the work they have done will stimulate similar courses in other prisons. More importantly, we hope that designers of future space stations and Mars bases will pick up some of the original ideas developed by the course participants and develop them to fruition.

Our plans are to continue this course with further Mars station design elements. At Glenochil prison, we have implemented a life support research program with prisoners interested in horticulture (the Mars Biopod project). This work will involve carrying out research on plant growth for the Moon and Mars in the prison greenhouses, allowing participants to get involved in real experiments to advance space exploration. As with the Mars station design work, the objective is for the participants to publish their work.

The *Life Beyond* course has helped harness the often latent talents of people in prison, helped generate positive contributions to society and stimulated hope for rehabilitation and future plans for life beyond prison.

We would like to thank Jim King at SPS, Head of Learning and Skills and Katharine Brash, Director Fife College (Prisons), who guided and supported the development of the course within SPS. We are very grateful to John Warttig and Peter Reilly at HMP Glenochil and

Edinburgh, respectively, who oversaw implementation and smooth running of the courses in collaboration with us at the University of Edinburgh and others in the learning centres at the respective prisons for their support. We thank the Governors of the prisons for their support in allowing for the implementation of the project.

I also would like to thank those at the University of Edinburgh who have taken part in this course, either in discussions about its progress or in teaching and implementation of the course. They include: Liam Perera and James Hitchen who taught on the first course, and Teun Vissers, Marialuisa Aliotta, Yair Fasado, Arthur Trew, Alan Shotter, Ana Azevedo-Da-Luz-Fialho and others who took part in the pilot scheme and provided useful ideas.

Above all, we thank the participants of this course, whose efforts, enthusiasm and hard work led to these new designs and ideas for the human exploration and settlement of Mars. They and we hope that this work will benefit the wider community of people working towards the construction of a space faring society.

INTRODUCTION:
FROM THE SCOTTISH PRISON SERVICE

James King,
Head of Learning & Skills, Scottish Prison Service
Katharine Brash,
Director, Fife College (Prisons)

We are both delighted and honoured to have been asked to contribute some thoughts and observations regarding the ongoing phenomenon that is 'Life Beyond' in Scottish prisons. It is only just over a year or so since Professor Cockell met with us to tentatively discuss possibilities of delivering introductory astrobiology lecturers at Scottish prisons to gauge the interest and responses to the endless postulates and possibilities of life beyond our own small planet.

The self-evident fact of this introduction is testament to the work of Professor Cockell and his staff in inspiring and sustaining engagement and the practical support services of our learning centre and operational staff in making this happen in Scottish prisons.

To enable the reader to better understand the context of education in prison it perhaps apposite to state that the Scottish Prison Service (SPS) in partnership with our core learning providers Fife College launched an innovative learning strategy in 2016. This document made clear our clear intentions: "to ensure that everyone in our care has the opportunity to engage in creative and flexible learning that unlocks potential, inspires change and builds individual strength".*

While this may appear as a standard corporate statement of intent it is actually quite radical in its intent of distancing custodial education from a fairly universal and uniform curriculum replete with remedial literacy/numeracy and low employability courses. In rejecting such 'correctional' pedagogy, SPS were both breaking with longstanding custodial conventions and realigning education to meet the individual needs and aspirations of learners.

---

*http://www.sps.gov.uk/Corporate/Publications/Publication-4017.aspx

Working closely with our long-standing partners, Fife College, these decisions are based on our experiences of what best engages reluctant and often disaffected learners. Our shared ambitions in in recent years have resulted in wide-ranging success through Project Themed Learning, Creative Arts and innovative partnerships with a growing number of Scottish universities in a network that is committed to assisting us in enrichening the education experience of those in our care.

This person-centred and social practise approach to individual educational opportunity has obvious limitations in terms of what can be delivered within a custodial context and within a public service challenged - like some many others - in an age of ongoing austerity. However, despite such challenges, the SPS have recognised and supported the transformative power of learning and firmly endorsed a liberal arts approach to education as the best means of cultivating latent talent, stimulating curiosity and transforming lives.

Initial responses from learners to the introduction of 'Life Beyond' in Scottish prisons varied between "what's astrobiology?" to mild scepticism or amused curiosity. However, it soon became apparent that the subject matter not only engaged learners but enabled them to imagine other entities and other realties and other possibilities. This helps learners to think more criticality, furnishing them with skills to re-imagine their own lives, appreciate the necessity and benefits of cooperation, citizenship and human endeavour and how they, as marginalised and often disaffected learners, could and should embrace the opportunities to contribute to the unfolding and flourishing of new knowledge and understanding.

The most recent development of initiating horticultural trials to inform possibilities of plant growth in space is testimony to the myriad of possibilities for ongoing academic cooperation and expansion of the boundaries of educational innovation and research.

Accordingly, we are very privileged and honoured to play a part in the unfolding of 'Life Beyond' and its ground-breaking research and creative staff. We look forward to considering future possibilities as we push the boundaries of educational endeavour and creative collaborations to transform the lives of learners.

LIFE BEYOND

FROM PRISON TO MARS

DESCRIPTION OF THE WORK

Charles Cockell

UK Centre for Astrobiology, School of Physics and Astronomy,
University of Edinburgh, Edinburgh, EH9 3JZ

The papers presented here are the products of the four-week course in which participants met with the Edinburgh team once a week for the duration of the course for two and a half hours at each sitting. The collective group effort in these papers therefore represents 10 hours of work. The participants had the chance to work on their own between the meetings. The reader should be aware that there was no opportunity to carry out detailed feasibility studies on segments on the work over the limited time available. But hopefully the reader will be impressed with the originality and ideas developed in such a limited time. Every effort has been made to remove factual inaccuracies and correct other errors, but otherwise the work has been left as it was presented to preserve its meaning and intentions. The groups were encouraged to develop new concepts and not to be overly concerned with technical detail given the four-week duration of the program and limited access to technical material. Artistic licence is therefore to be found in the designs, but we hope the reader will find the concepts and ideas exciting and an inspiration to develop further.

The work is divided into the Mars station concepts from HMP (Her Majesty's Prison) Glenochil and Edinburgh. There are two station concepts from each prison. We also present the contributions to the *Mars Project Creative Writing Competition* that was implemented at HMP Glenochil during the course.

HMP Glenochil Mars Station Designs

Fifty Years on: Development of Mars Terra Nova Station

The Fifty Years on project was the work of a group at HMP Glenochil. The project envisages the state of Mars exploration fifty years after the first stations have been constructed. It describes a scenario of two bases on Mars, one near the north polar ice cap and one in Meridiani Planum, and an orbital infrastructure. All of these facilities are linked together into

1

a coordinated Mars settlement architecture. The concept is presented as a single comprehensive paper.

Elysium Station

The Elysium station project was the work of a second group at Glenochil. The focus of their work was to design a base for the near-term on Mars and then to consider the development of Mars over the next two hundred years at a point when the original base has become a museum in a much-expanded station, now focused on tourism. The work is presented as three separate short papers:

1 Initial Elysium Station and Rationale

This paper can be regarded as a summary document on the first Elysium station to be established on Mars. The document expresses some of the reasons for setting up the station and whether humans or robots should be used to build the initial base. The document lays out the concept for three initial missions to Mars to establish the station before further growth, and the focus of those missions. The document also presents concept drawings for the initial Elysium Station and its structure and design.

2 Elysium Holiday Complex

Two hundred years in and Elysium Station has been transformed into a holiday and tourism complex. This paper presents some of the basic specification and an art illustration of the new station in 2233. A poster to entice Earth-bound tourists to visit Mars and that might adorn travel agencies on Earth in the 23rd century is shown.

3 Mars – A Brief History of the Colonisation of the Red Planet and the Development of Elysium Station

This paper presents a fictitious time line of international and interplanetary events over the 200-year history of the station (2030-2233) with some of the major events that shape the development of the station. The paper can be regarded as a work of science fiction, but at the same time, it provided a structure to the thoughts of the group by generating a time line that guided discussions. Some of the events that occur, particularly in terms of international cooperation, may even have real counterparts in the years to come. The document lays out the major events occurring with Elysium Station in the context of international and interplanetary developments.

HMP Edinburgh Mars Station Design

The Mars station design from HMP Edinburgh is presented as a single concept paper. **Mars Station – The Edinburgh Concept**

The Edinburgh Concept considers standardised modules that can be modified for a diversity of uses. The reasons for establishing a station on Mars are investigated and many facets of the station, from tourism to scientific exploration, are considered.

Mars Project Writing Contest

During the course, an ancillary activity was organised by Alan MacFarlane at the Learning Centre at HMP Glenochil. The *Mars Project Creative Writing Competition* had a simple remit: 'You are working on the Mars space station and you are writing your first email home. In no more than 500 words, what do you have to say?' The entries to this competition are reprinted here and display some of the imagination and ideas that the project stimulated.

FIFTY YEARS ON: DEVELOPMENT OF MARS STATION

John, Neil, Taylor, Martin, Gordon, John,
HMP Glenochil

I am convinced that humans need to leave Earth and make a new home on another planet. We need to do it now before humanity is overtaken by some disaster that we can neither anticipate nor control.

It will take more than explorers to colonise a planet. To build a self-sustaining colony, astronauts would need to build habitats, farm food and mine and process the planet's resources.

Perhaps robots would prep the location for the arrival of the first settlers. Establishing a colony in space would require incredible ingenuity, but ingenuity is a resource humans have in abundance.

Our species' natural curiosity is what will drive us to these distant planets. In the next 100 years, we will embark on our greatest ever adventure. Our destiny is in the stars.

Professor Stephen Hawking,1942-2018

Abstract

Fifty years after the first manned landing on Mars, what form will the principal human outposts there take? Supporting 160 crewmembers, Mars Station as envisaged in this paper will consist of two planetary bases connected to an orbital space station (SkyStation) by a Space Elevator. The principal base will be located in the upper northern latitudes of the Vastitas Borealis, close to the polar ice-cap; and the secondary base will be located on the equator at the Meridiani Planum.

Base facilities include power generation, manufacturing, aquaculture and laboratory sites in addition to residential, communal and medical facilities. As far as possible, Mars will have become self-sufficient.

This paper focuses on potential solutions to the technical challenges that will be encountered in the colonisation of Mars, including the generation of oxygen, water, food production and power generation. For food production, the use of aeroponics and cultured protein is recommended. Power generation options are considered, including renewables as well as nuclear, with fusion – in particular the CrossFire reactor model – being preferred. Radioisotope Power Systems are discussed.

Transportation systems will be been developed for rapid movement between the bases using Hyperloop technology in addition to the deployment of Semi-Rigid Dirigibles, Unmanned Aerial Vehicles, Multi-Purpose Vehicles and other systems.

In order to minimise risk to humans, much of the exploration and exploitation of Mars will be carried out by robots in conjunction with immersive VR systems.

This paper also provides a brief overview of Martian governance. In order to avoid commercial or national conflict, the planet will be operated by a monopoly corporation equally owned by all interested parties.

Keywords: Mars, fusion reactor, aquaponics, aeroponics, Hyperloop, plasma-catalytic process, areostationary satellites, robotics

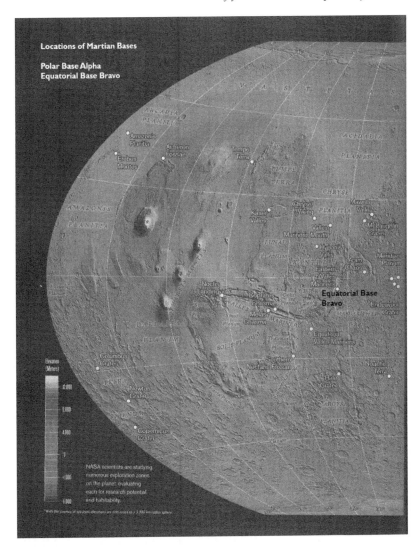

Figure 1a. *Map of Mars showing Bravo (Equatorial) Base*

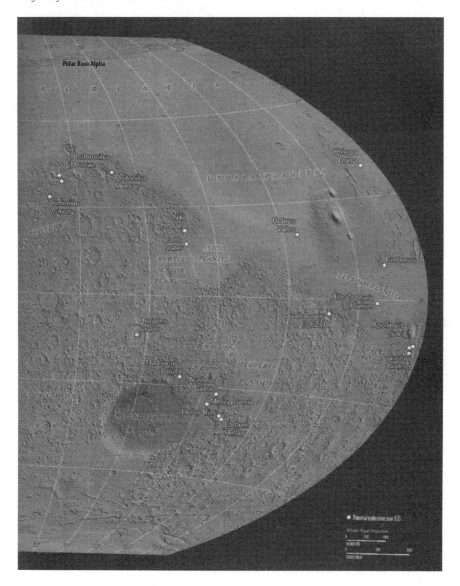

Figure 1b. *Map of Mars showing Alpha (Polar) Base*

1 Project Scope

The scope of our project **Life Beyond: Project Mars** was determined through a combination of the guidance and parameters set by Professor Cockell and his associates from Edinburgh University and our own group discussions.

These relate to the challenges and potential solutions of developing a human colony on Mars, with the base parameter being that it is set some fifty to seventy five years after the first manned landings: effectively projecting forward a century from now.

This team was one of several in HMP Edinburgh and HMP Glenochil which were involved in the overall project, with each taking a different perspective on the challenges facing human settlement and exploration of Mars.

Team members assumed responsibility for elements of the project and investigated specific challenges in order to come up with solutions. These were brought together through group discussions to form the proposals contained within this document.

1.1 Considerations

Time frame
- Near to mid-future placing the colony within 50-75 years from the first manned landing

Technology
- All solutions will be grounded within current and emerging technologies as well as hypothetical engineering solutions based on those technologies.

Financial
- No restrictions or cost factors were taken into account when considering solutions to the challenges identified within this project. This allowed the team to apply 'Red Sky' thinking and push the boundaries of potential solutions.

Parameters

- The presumption has been made that the colony is already established and we have not concerned ourselves with previous expeditionary settlements or expeditions.

Logistics

- This project is not considering the transportation of persons or materials from Earth to Mars or any vehicles which would be involved in those processes. Logistics are confined to Martian planetary arrangements.

1.2 Goals and Objectives

We decided on the purpose and objectives for our colony following group discussions which determined the following:

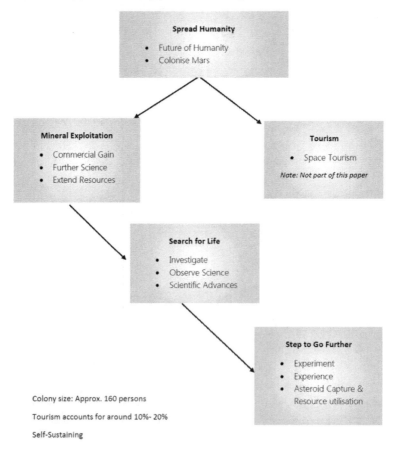

2 Life Beyond

2.1 Challenges and solutions of setting up a colony on Mars

Challenges	General Solutions	Specific Solutions
Human Factors Conservation Biology	• Screening required & selection of colonists to be balanced (where long term family units are to be selected)	◇ 500 adults required to ensure a typical evolutionary lifespan. ◇ Optimum size for colonisation would be the size of a small village - approx. 160 pax (John Moore, anthropologist: University of Florida). (1)
Genetic Isolation / Drift	• Genetic screening to ensure genetic biodiversity • Genetic engineering to be considered	◇ Genetic Base Editing to prevent defects and disease (distance and isolation to offset moral / ethical issues). ◇ Biohacking with sensors / communication devices and potential other transhuman technology.
Psychological Effects	• Psychological profiling and testing of any prospected colonists • Meaningful and purposeful living and working	◇ Psychological distortions to be minimised by using normal life rhythms and use of Zulu time. ◇ Mental Health support to be provided on site and via communication with Earth. ◇ Part of larger project to launch missions to asteroids such as Memusa / Chiron for resource collection. ◇ Use of VR/ MR and Augmented reality systems to reduce issues associated with being 'off world'. ◇ Domes and surface activities to supplement habitat work to help minimise confinement issues.
Physiological Effects	• Reduce confinement by expanding and diversifying habitat. • Use of physical activity to ensure strength of bones and reduce the risk of muscle atrophy	◇ Implementation of exoskeleton technology to augment capabilities of colonists.
Law / Society	• Article II of the 1967 Outer Space Treaty: *No nation has sovereignty or control of any satellite bodies.* This treaty refers to the Moon and has no mention of individual rights or corporate rights.	◇ Update of treaty with issues of private and corporate ownership and rights to be addressed. ◇ Colonial law or rules to be drawn up and agreed by all.

Challenges	General Solutions	Specific Solutions
Martian Environment No atmospheric Oxygen (0.15%)	• Atmospheric CO_2 could be utilised in order to extract the Oxygen • Use the CO_2 frozen ice cap at the Southern Pole to extract Oxygen • Extract Oxygen from the H_2O ice cap at the Northern Pole or through the permafrost • Growing of plants to convert CO_2 into Oxygen through photosynthesis	○ Air-source capture and extraction system ○ Low temperature plasma decomposition of CO_2 from atmospheric / frozen sources ○ Electrolysis to extract the Oxygen from H_2O ice ○ Nanoclimites or artificial plants to convert CO_2 into Oxygen. ○ Use of micro bacterial digestion to convert CO_2 in atmosphere or CO_2 ices into Oxygen
Radiation High doses of radiation due to lack of magnetosphere	• 1 millisievert / day on the surface – 180 day travel to Mars = 500 millisievert against a background radiation on Earth of approx. 5 millisievert per year) • Protection from Ultraviolet rays / Cosmic rays could be provided by having a habitat which was subterranean • Bring radiation shielding materials from earth • Build habitats in polar ice regions or utilise permafrost as a shield against radiation • Use cave(s), valleys and natural landscape to assist in shielding from radiation	○ Hybrid habitat solution building the structure of the habitat both above ground and subterranean with insulating to limit exposure. ○ Water and ice could be considered as shielding from radiation ○ Arens such as the Valles Marineris could be used to benefit from natural shielding ○ Utilise lava tubes in areas of past geological activity such as the Tharsis Bulge (Pavonis Mons / Arsia Mons) (2)
No Magnetosphere Meteors / Micro-meteors	• Build habitat(s) / modules from locally sourced materials which can be supplemented with underground chambers to protect from impacts	○ Construct habitat using a mixture of silica and iron materials turned into ceramic via the use of microwaves and use as protection ○ Mavroidisbots – Protein dispensing nanobots secrete proteins to seal structures in the event of punctures *(Prof Konstantinos Mavroidus NEU)* (3)
Low Atmospheric Pressure (0.6 – 0.8 kPa) Cup of water would evaporate in seconds	• Pressurised habitats / suits and vehicles required due to low atmospheric pressure. *NB: Without the use of terraforming solution such as melting CO_2 ice to provide additional pressure* • Robots / Co-bots impervious to pressure issues to be utilised to overcome challenges • Bio-domes to protect colonists and crops from low atmospheric pressure	○ Next generation space suits with new materials and jointing. ○ Use of Telepresence devices for prolonged surface operations. ○ Valkyrie R5 type design electrically powered humanoid

Challenges	General Solutions	Specific Solutions
Low gravity 37% of that on Earth	• Use of physical activity and restraining straps to provide extra force in order to combat effects of low gravity • Gravity simulated by spinning elements of colony in a centrifuge type arrangement	○ Microgravity could be countered by colonists rotating into an orbital vehicle (SkyStation) where a counterweight was employed to generate a spin of around 2rpm (based on a 1.5km tether) to simulate Earth Gravity SkyStation to be linked to orbital ascension and descending vehicle (Space elevator)
Low sunlight levels 50% - 33% of the sunlight on Earth	• Geodesic domes utilised to expose plants and colonists to natural sunlight • Enhanced sunlight (0.33-0.5 of sunlight on Earth) via sun mirror	○ Areostationary orbital platform used to capture and focus sunlight (sun mirror) in order to beam it to the surface. Satellite to be large in construction and constructed of an aluminised Mylar material. *(Chris McKay NASA astrobiologist)*
Temperature Ranges from high of 21°C at equator in summer to a low of -100°C in winter at the poles. Average temperature = -27°C	• Insulated habitat(s) utilise locally sourced materials and heated via microwaves to provide insulation • Choice of location of habitat to minimise temperature fluctuations arising from seasons and orbital eccentricity • Location of Habitat to utilise possible geothermic activity • Use of solar heat exchangers to produce heat • Ground source heat pump solutions via bore hole	○ Heating of habitat via potential geothermic means. Areas where deep formed liquid magma may still be present and yet not in an explosive nature (due to low gravity) such as Tharsis Bulge and Elysium region could be utilised ○ Recycling of heat from processes such as power generation and confinement of HT Plasma ○ Use of inward reflective glazing materials and solar luminescent salt generating processes to aid with the conservation and generation of heat for the habitat
Dust storms (100km/h) Particulate size 1.6 ± 0.5µm and magnetic composition	• Habitat to be positioned in sheltered position potentially within the lee of terrain and in a combination of surface and subterranean units. • Shuttering / louvres and protection against dust storms to be used on exposed elements such as Geodesic Domes. • Filtration systems required to protect atmospheric processing facilities	○ Utilise wind power through deployment of turbines to power part of the habitat ○ Use of 'Smart Motes' or larger robotic autonomous devices to clear the dirt and debris from domes and solar arrays ○ Magnetic separation of dust particulates to protect both base environment and equipment
No Liquid Water	• Extract liquid from H_2O ices in the Northern polar region/ permafrost • Use potential subterranean aquifers where water may be kept liquid by radioactive decay and pressure from rocks above • *Water on Mars M H Carr*	○ Heating of habitat via potential Geothermic means. Areas where deep formed liquid Magma may be present and yet not in explosive nature (due to low gravity) such as Tharsis Bulge and Elysium region could be utilised ○ Use of Solar furnaces to focus and utilise sunlight. ○ Polar inventory of water ice shows approx. 3.2-4.9 million km[3]

13

Challenges	General Solutions	Specific Solutions
	• Extract H_2O from atmosphere (limited resource) • Areas showing signs of thermokarst to be investigated for permafrost / trapped H_2O • Solar Power Reactor to turn regolith into water and oxygen • Process using Zeolite to extract water from the atmosphere	o Use of Planetary Spectroscopy to find sources of potential subsurface water o Areas such as Deuteronilus Mensae have signs of potential glaciated deposits seeping from valley ends. Water recyclers to be used to limit the wastage of resources o Potential use of renewable power system (see power section) to generate H_2O from atmospheric CO_2 o (Thorsten Denk – SolarPACES conference 2017) (4) o NASA development programme to remove moisture from the atmosphere (nearly 100% humidity). Pass the atmosphere over the mineral Zeolite which absorbs the moisture which in turn can be removed as water
Communications Distance 10-40 mins latency period (depending on orbital positions) for communications with Earth	• Communications array set up on the surface for communication • Orbital satellite relay stations to be launched to allow for global communications and for continuous communications to Earth • Command and communications satellites to be put into orbit to overcome the LoS communications issues	o Low Mars Orbit (LMO) for relay of digital communications o GPS system through use of positioning satellites o Areostationary orbiting satellites at equatorial orbit of 17032km o HDR and LDR links
Power / Energy Source	• Solar Power – Photovoltaic cells in a Solar Array • Solar Lenses / Solar Mirror • Solar converting windows and domes • Wind Turbines to supplement power when wind storms are active • Fusion Reactor He_3 • Fusion Reactor – Deuterium / Tritium • Conventional PWR Fission reactor to generate heat to be transferred into electrical energy • Hydrogen Cell(s) • Geothermal Power	o Use of Solar Mirror in orbit to deflect and focus sunlight onto the Solar Array on the surface o Use impregnated windows and domes with transparent solar luminescent salts which convert sunlight into electrical power o Use of both wind and solar energy to power CO_2 processing plant. CO_2 extracted from atmosphere and cooled / compressed using 'renewable energy'. When no wind or at night CO_2 can be used through a turbine to generate power and recaptured at the other end to be used for CO_2 splitting into oxygen. o He^3 to be harvested locally to power a He^3 Fusion Reactor. He^3 is high yield approx. 107 million homes could be powered for 1 year on 6 000kg of He^3.

Challenges	General Solutions	Specific Solutions
	• Positive Ion Power Capture from atmosphere, including in particular dust storms • Positive Ion Capture from Phobos • Transmission of power over microwave power transmission grid	○ Hydrogen isotopes could be used to provide power through a Fusion Reactor giving 10^8MJ/kg of energy. Tokamak device which confines hot plasma at around 100m^{3}K ○ Fission Reactor – conventional nuclear power reactor to generate heat – turbine power in a closed water system Integrated Hydrogen fuel cell power to provide electrical power for vehicles etc. ○ Possible use of Geothermic Power – turn water – steam in conventional heat exchange process. Dependent on location and access to magma ○ Positive Ion capture utilising atmospheric capture devices with power being transmitted over microwave grid network ○ Utilisation of the solar wind at 1.6 million km/h on the surface of Phobos and beaming the generated power over MW grid. Due to wake effect caused by Phobos the Ions and electrons travel at differing speeds forcing the ions into a plasma void behind the leading face of Phobos. *(William Farrell – Goddard Space Flight Centre) (5)* ○ Use of Microwave technology to transmit power
Habitat Martian Planetary Environment	• Modular habitat sent in advance of colonists and constructed by AI robots such as the R5 Valkyrie • Modular ship built and sent to Mars which 'transforms' on the surface to become a multifunctional habitat • Inflatable domes used to provide temporary accommodation – CO_2 from extracted and compressed from atmosphere. Construction linked to 3D printer to produce habitat. • 3D printing of habitat from local materials • Use of robotics + AI to fabricate habitat(s) or hybrid surface/subsurface habitats	○ Use of honeycomb fibre MS materials (3cm x 0.02mm can support up to 2 000kg) ○ Use of linicube technology (105 ways pf plugging one component into another) to construct habitats ○ Building using tessellating triangles which can form almost any shape. Manufactured on-site using solar furnaces/3D printers + local materials to produce fine, light + strong silicon components ○ Use of graphene – highly conductive to repel radiation and for structural strength ○ Constructing habitat (or parts thereof) from carbyne nanotubes knitted together

Challenges	General Solutions	Specific Solutions
Resupply of habitat	• Orbital facility or station to facilitate resupply of colony and for imported materials for habitat • Tether or 'Space Elevator' to be constructed with capsule or 'cars' to allow for the transportation from the surface to the orbital facility	○ Silicon + iron bearing minerals can be fused into solid ceramic building materials using microwaves (NASA: Aerospace Scholars Program) (6) ○ Construction of a Space Elevator extending into areostationary orbit near the habitat. Design could incorporate elements such as Fullerenes (carbon nanotubes and molecular carbon materials) as proposed in the design study *Space Elevators by P. Swan et al.* (7)
Death of colonist and general disposal / recycling of organic remains	• Use of Alkaline Hydrolysis in the form of a Resomator	○ Resomator to use Potassium Hydroxide and water heated to 150°C for complete breakdown of organic matter which could then be recycled into the colony bio-systems
Escape Plan	• Use of orbital facility to enable the evacuation of some or all of the colony in the event of an emergency	○ Orbital facility could be either a holding centre for evacuation to Earth or its own self-contained vessel for transport back to Earth
Discovery of Life on Mars Microbes in subsurface aquifers / oasis (possible biogeochemical model for Mars by A De Morais 2012)	• Isolate and prevent contamination or cross contamination of potential • Decontamination protocols for colony / colonists which may come into contact with Martian materials or contaminants • Existence of extremophile organisms	
Food Sustainable source / supply of food Resupply from Earth	• Use of Martian soil combined with fertilisers and microbial material for the production of crops • Construct Bio-domes to grow crops / food stuffs in order to feed colony (enhance sunlight with solar mirror) • Use hydroponic systems for the production of food • Grow proteins and 'artificial meat' in the lab • Resupply colony from earth with shuttle runs to an orbital facility	○ Genetically modified agriculture specifically engineered to grow in environment

3 Human Factors

3.1 Working schedule

Taking into consideration the six month transit time between Earth and Mars, it will be necessary for all personnel working there to be on a minimum three year assignment [8].

The planetary work schedule will be on a four month rotation, consisting of one month on the SkyStation in Earth-like gravity (one G) conditions in order to minimise the adverse neurological physiological effects of reduced gravity. The other three months will be on assignment to one of: Polar Base Alpha, Equatorial Base Bravo or an exploratory mission.

The working week will be the same as on Earth: five on, two off; with an eight hour working day.

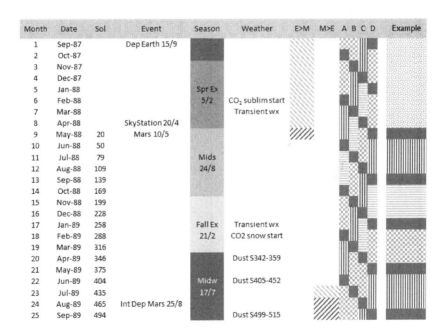

Month	Date	Sol	Event	Season	Weather	E>M	M>E	A	B	C	D	Example
1	Sep-87		Dep Earth 15/9									
2	Oct-87											
3	Nov-87											
4	Dec-87											
5	Jan-88			Spr Ex								
6	Feb-88			5/2	CO_2 sublim start							
7	Mar-88				Transient wx							
8	Apr-88		SkyStation 20/4									
9	May-88	20	Mars 10/5									
10	Jun-88	50										
11	Jul-88	79		Mids								
12	Aug-88	109		24/8								
13	Sep-88	139										
14	Oct-88	169										
15	Nov-88	199										
16	Dec-88	228										
17	Jan-89	258		Fall Ex	Transient wx							
18	Feb-89	288		21/2	CO2 snow start							
19	Mar-89	316										
20	Apr-89	346			Dust S342-359							
21	May-89	375										
22	Jun-89	404		Midw	Dust S405-452							
23	Jul-89	435		17/7								
24	Aug-89	465	Int Dep Mars 25/8									
25	Sep-89	494			Dust S499-515							

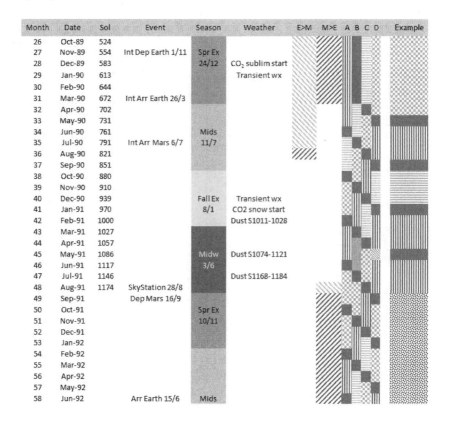

Month	Date	Sol	Event	Season	Weather	E>M	M>E	A	B	C	D	Example
26	Oct-89	524										
27	Nov-89	554	Int Dep Earth 1/11	Spr Ex								
28	Dec-89	583		24/12	CO₂ sublim start							
29	Jan-90	613			Transient wx							
30	Feb-90	644										
31	Mar-90	672	Int Arr Earth 26/3									
32	Apr-90	702										
33	May-90	731										
34	Jun-90	761		Mids								
35	Jul-90	791	Int Arr Mars 6/7	11/7								
36	Aug-90	821										
37	Sep-90	851										
38	Oct-90	880										
39	Nov-90	910										
40	Dec-90	939		Fall Ex	Transient wx							
41	Jan-91	970		8/1	CO2 snow start							
42	Feb-91	1000			Dust S1011-1028							
43	Mar-91	1027										
44	Apr-91	1057										
45	May-91	1086		Midw	Dust S1074-1121							
46	Jun-91	1117		3/6								
47	Jul-91	1146			Dust S1168-1184							
48	Aug-91	1174	SkyStation 28/8									
49	Sep-91		Dep Mars 16/9									
50	Oct-91			Spr Ex								
51	Nov-91			10/11								
52	Dec-91											
53	Jan-92											
54	Feb-92											
55	Mar-92											
56	Apr-92											
57	May-92											
58	Jun-92		Arr Earth 15/6	Mids								

Key

	Mars		Earth	
	Sols	Months	Days	Months
Spring	194	7	200	7
Summer	177	6	182	6
Autumn	142	5	146	5
Winter	156	6	160	5
TOTAL	669	24	688	23

1 Sol – 24 h 39 m 35 s

1 month = 28 Sols

- Transit to Mars
- Transit to Earth
- Alpha
- Bravo
- SkyStation
- Exploration
- Earth > Mars
- Mars > Earth

3.2 Martian climate

Figures 2 and 3 illustrate the significant differences in temperature between the northern and southern hemisphere summers: the northern being cooler than the southern.

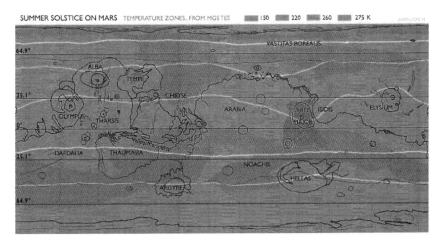

Figure 2.Temperatures on Mars at Summer Solstice

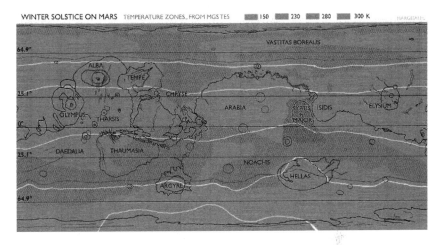

Figure 3. Temperatures on Mars at Winter Solstice

The Martian climate poses challenges for exploratory operations, in particular during dust storm periods. Temperatures at the Polar Base are likely to cause operational difficulties, in particular during winter in the northern hemisphere.

There are significant levels of atmospheric dust on Mars, much of which is ferrous. Atmospheric heating associated with this generates the ubiquitous Aeolian activity on Mars. This varies from near-constant 'dust devils' through to global dust storms occurring every three or four Martian years.

Atmospheric H_2O in the form of vapour and ice clouds has a significant role in atmospheric chemistry, dust-density and climate.

The ferrous dust is also likely to contribute to significant electrical activity through charged particles and their discharge. Dust adhesion may also be affected by electro-magnetic electrical effects, which is likely to have a significant impact on operations including in particular communications.

4 Habitat

4.1 Selection of Base Locations

The principal base (Alpha) will be located within the northern Vastitas Borealis polar region, close to the ice cap. The northern ice cap consists of water and its exploration will be a primary long term research goal, achieved by drilling for core samples [9].

Polar ice will be converted to potable, demineralised and other water supplies and used to operate the fusion reactor as well as providing oxygen and hydrogen through electrolysis.

The location of the secondary base (Bravo) is predicated by its primary function as the arrival port for Mars, connected to the areosynchronous SkyStation by means of a tether used by the Space Elevator. Given this requirement, it is logical for it to be located in close proximity to sites which are of exploratory interest: hence its location at Meridiani Planum.

Figure 4. Meridiani Planum

As far as possible, habitats will be constructed from materials mined on asteroids and locally on Mars, thereby minimising the requirement to ship materials from earth.

4.2 Polar Base Alpha

Figure 5.Martian North Polar Cap

Polar Base Alpha will be the principal base on Mars. Its location in close proximity to the northern polar icecap has been selected to enable easy access to ice that can be turned back into water as well as providing core samples for scientific research purposes.

The general configuration of the base will be a 'snowflake' with spokes coming off a central hub.

The vast majority of functions on Mars will be automated, and the control centre will be the hub of operations. This will include stations for the VR control of UAVs, vehicles and robots. It will also house links to the satellite constellation (see Section 7) enabling terrestrial and interplanetary communications including with the SkyStation.

External access will be through Extra Base Activity (EBA) prep-room, airlocks and decontamination for Rover/vehicles and Semi-Rigid Dirigible (SRD) load/offload hangar. All will utilise separate 'in' and 'out' airlocks/channels to ensure biosecurity.

Workshops will be provided for the construction and repair of SRDs, satellites and robots.

Accommodation will consist of units providing galley/catering, recreation and sleep modules. The recreation modules will include exercise facilities with electronic gaming/VR. Sleep modules will consist of individual and shared rooms with sanitation including toilets, showers and washing facilities. Laundry facilities will also be available.

Activities include the production of food crops and protein, the assembly and repair of semi-rigid dirigibles, satellites and robots [10].

Food production will take the form of fruit and vegetable crops and protein. The crops will be grown in a range of environments including hydroponics beds, aquaponics systems, and soil cultivation. Protein will be derived from insects and artificial meat.

Solar, wind and nuclear energy sources will provide power as well as essential life support system (LSS) gases and water. Atmosphere management in the base will include O_2 production and supply together with CO_2 scrubbers. The air conditioning and pressurisation system will enable dust prevention and removal through magnetic filtration. Water will be sourced from the northern polar ice cap and will be filtered and purified by exposure to UV light. Demineralised water will be produced for the nuclear powerplant. Other LSS systems will include waste processing and fire detection and suppression.

There may be residual geothermal activity which could be tapped to provide heating through ground source exchange.

Principal research activities carried out from Polar Base Alpha will include climatology, glaciology and astrobiology.

The scientific facility will include a human biology laboratory linked to the medical suite and a Martian geological and biological laboratory. This will carry out decontamination and packing of samples to be sent to Earth.

The Medical Trauma Centre (MTC) will include an MRI/body scanner together with comprehensive ophthalmic and audiology testing facilities, an Obstetrics and Gynaecological (ObGyn) centre including perinatal facilities. It will include a human biology laboratory which will test water and air quality together with possible contamination of food preparation/production areas and general accommodation on a continuous basis. It will monitor personnel for biohazards and radiation exposure and for testing and modification of DNA. Dental and physio treatment rooms will be provided. Pharmaceuticals will either be shipped from Earth or manufactured locally.

The surgical suite will include a recovery room/ICU. In order to avoid contamination, blood will either be taken and stored for the exclusive use of the person providing it, or alternatively artificial blood will be created on site. There will also be a VR surgical control point for all Air Transportable Surgical Unit (ATSU), which will be held at both Polar Base Alpha and Equatorial Base Bravo for closest despatch.

Rigorous psychological screening and testing will be deployed for prior to, during and subsequent to deployment to Mars. Cognitive behavioural therapy and other psychological treatment including relaxation and de-stressing will primarily be delivered through VR devices.

A Resomator (alkaline hydrolysis) will be used for disposal of bodies and medical waste.

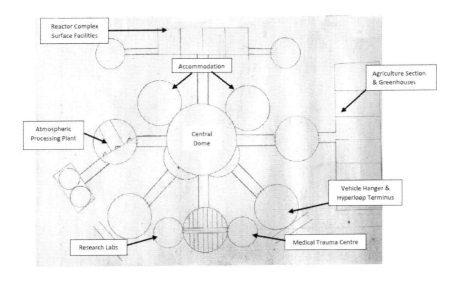

Figure 6. *A diagrammatic view of Polar Base Alpha*

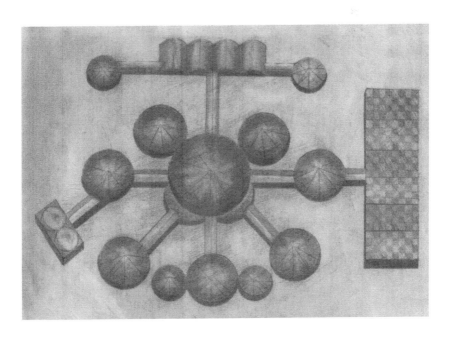

Figure 7. Artist's impression of Polar Base Alpha

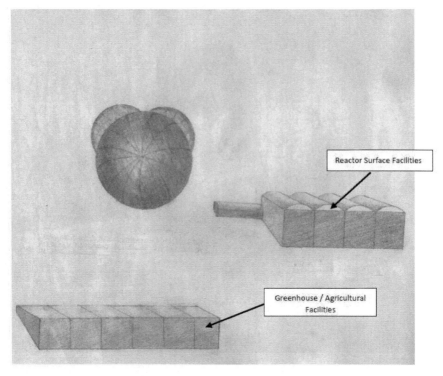

Figure 8. Surface View of Polar Base Alpha buildings

These represent the structure of the surface buildings which would be constructed where possible from local in situ resources.

4.3 Equatorial Base Bravo

Equatorial Base Bravo will be located at the Meridiani Planum and is the main point of arrival on Mars, for people, equipment and supplies and as such is the planet's primary logistics centre. The base is responsible for handling, breaking down and distribution of cargoes to Polar Base Alpha primarily via the Hyperloop transport system and to remote exploratory sites as required, using surface vehicles or SRD's.

The base also handles all of the decontamination and shipment of materials, equipment and rotable spares as well personnel for transport back to Earth. These operations will be carried out via the docking system located on the orbital SkyStation. The base will also contain a Launchpad for uplifting of heavier loads from the Martian surface for RTE transfer or for emergency transport to the SkyStation.

Figure 9. *Location of Meridiani Planum*

Similarly, this base is where outgoing astronauts are quarantined and checked for any infections or viruses before leaving Mars.

Equatorial Base Bravo will be the principal base for all exploration of the equatorial and southern hemisphere regions of Mars.

Some of the specific facilities incorporated into the Equatorial Base Bravo are detailed below:

Space Elevator Dock
This is the arrival and departure point for persons and cargo to and from the SkyStation and therefore Earth. The dock consists of an elevated structure built around the surface anchor for the elevator tether and cable which has Separate 'in' and 'out' airlocks/channels to ensure biosecurity and to enable quarantine of persons or cargo which may be deemed hazardous.

27

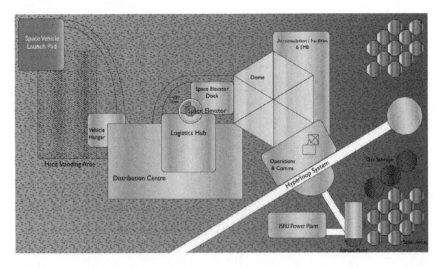

Figure 10. *Key features of Equatorial Base Bravo*

The dock for cargo (up to 20 000kg) is managed and operated by robotic devices and is segregated from the passenger terminal to ensure safety and maintain biosecurity.

The passenger terminal leads to the main transportation 'car' which is the vehicle for transportation up the 17 000km cable to the SkyStation. The control and operation room for the vehicle is separate to the dock facility as the system is automatically operated.

Emergency Medical Bay (EMB)
Attached to the Space Elevator dock will be the Emergency Medical Bay as well as the suite for the screening of persons entering and leaving the Martian surface. The EMB will be fully equipped to respond to emergency situations and be staffed by Advance Medics and Advance Nurse Practitioners and where necessary Doctors (see Section 11). These staff will be assisted by robotic devices and by virtual remote assistance from the main Medical Trauma Centre located at Polar Base Alpha.

The EMB will also house the equipment for carrying out medical checks and quarantine of outgoing astronauts as well as decontamination facilities and a small quarantine / isolation unit.

On standby at the Base will be the remotely operated Air Transportable Surgical Unit (ATSU) which can be dispatched to any incident to recover or evacuate casualties to the EMB or for more serious conditions to the Medical Trauma Centre.

Logistics Hub
The logistics hub will be for the storage, transfer and distribution of all materials both on and around the surface and to off-world destinations. The centre will be linked to the cargo dock and space elevator dock as well as the space vehicle transfer launching platform.

The hub will be attached to the logistics distribution centre which will be segregated into incoming and outgoing channels (as with the passenger terminal) and be controlled by an Artificial General Intelligence system. This AGI system will direct and task robotic devices to handle the movement and distribution of supplies, cargo, products and materials.

The logistics centre will be connected to the Hyperloop transportation system to allow for the quick and easy transfer of items to the main Polar Base Alpha (see Section 8.1) and also to the transportation hub where cargo can be loaded onto the surface vehicles or SRD's for transfer to remote locations on the surface.

Decontamination and quarantine of outgoing geological and biological samples / products (to remove potential bio-contaminants) will be carried out within the logistics centre.

General Accommodation and Facilities
The general accommodation will be similar to those at Polar Base Alpha however more compact and on a smaller scale with less facilities for recreation. The main human habitation and transit accommodation will consist of:
- Accommodation and sleeping quarters
- Galley/catering
- Showers/washing facilities
- Exercise / gym facilities
- Recreation
- Library / Electronic gaming/VR

Life Support Systems (LSS)

Life support systems are on a smaller scale than those situated at Polar Base Alpha, however the provision of power and atmosphere is still required to run on similar models. It is envisaged that the location of the base lends itself to the use and utilisation of renewable energy sources such as solar and wind (Section 6.4.1) and these will be the primary sources of power which will be supplemented by a CrossFire fusion reactor (Section 6.4.3). The LSS will have many standard components required to support human life such as:

- Atmosphere management (Section 6.1)
    - O_2 production and supply
    - CO_2 scrubbing
    - Air conditioning and pressurisation
    - Electrostatic / magnetic filtration
- Temperature control
    - Ground Source Heat Recovery
    - Electrically generated heating systems
- Water / waste management (section6.2)
    - Recycling and purification / demineralisation
    - Waste processing

5 Extra-planetary Infrastructure

5.1 SkyStation

After the 33-million kilometre journey from Earth taking approximately six months, the first sight of the SkyStation will be eagerly awaited by those on board the Interplanetary Shuttle. Looking somewhat like an enormous doughnut floating above the surface of Mars, the automatic Shuttle is guided in by a pilot on the SkyStation using VR to a dock on the outer ring.

When the weak rays of the sun shine on it, the tether for the Space Elevator has the appearance of a beam of laser light rising from the ground through the middle to the SkyStation and onwards into space.

Passengers disembark and are processed through comprehensive health checks to ensure no infections or viruses are brought onto Mars. All incoming astronauts are quarantined until it is certain they are healthy, after which they will undergo an induction process prior to travel to Mars.

This will include their designation as members of one of the four teams. This determines their duties – each team is rotated between the three bases (Polar Base Alpha, Equatorial Base Bravo and the SkyStation) in addition to additional roles depending on specialisation and conditions.

There is strict segregation between incoming and outgoing astronauts as well as those manning the SkyStation in order to ensure biosecurity.

The SkyStation will consist of a central complex through which the Space Elevator will dock, and a toroidal outer rotating ring or 'donut' which will have a permanent gravity equivalent to that of Earth. All astronauts work there one month in four in order to derive significant health benefits by preventing brain and muscular atrophy caused by low or zero gravity environments.

Incoming cargo should have been decontaminated prior to departure from Earth, but this will be double-checked to ensure there is no biological contamination prior to dispatch to Mars.

Travel between the SkyStation and Mars is on the Space Elevator to Equatorial Base Bravo.

The SkyStation will have an additional role as a 'lifeboat' in the event of an environmental or catastrophic event on Mars which requires the evacuation of astronauts from the surface. It will be equipped with long-range ion-powered vehicles that can be used to transport settlers back to Earth or another planet.

- Cargo
    - Incoming goods should have been decontaminated prior to departure
    - QC checks should be made to confirm this and no alien microbes/bio-organisms or seeds contained within shipments
- Medical checks and quarantine for incoming astronauts
    - Health centre
        - Laboratory
        - Isolation rooms
- Space Elevator dock (pax + cargo handling)
    - Transit point for pax and cargo between Earth and Mars
    - Separate 'in' and 'out' airlocks/channels to ensure biosecurity
        - Tourists identified and escorted at all times
- Satellite launch platform
- Interplanetary communications relay
- Accommodation for SkyStation Crew
    - Galley/catering
    - Recreation
        - Exercise
        - Library
        - Electronic gaming/VR
    - Shower/washing facilities
        - Toilets
    - Sleeping quarters

- Life Support Systems
  - Power
  - Atmosphere management
    - O_2 production and supply
    - CO_2 scrubbing
    - Air conditioning and pressurisation
  - Water
    - recycling
    - potable supply
      - Filtration/purification (UV)
  - Waste processing
  - Fire detection and suppression

5.2 Space Elevator

The SkyStation will be connected in areosynchronous orbit by a space elevator. This will allow the cheap and fast movement and transfer of goods, material and personnel from the surface of the planet to low orbit without the need for resource depleting and environmentally damaging propulsion systems.

The elevator itself will consist of a surface anchored tether constructed from a combination of Carbyne Carbon allotropes and diamond nanothreads woven together [11]. This will extend in total to 34 000km with a counterweight positioned at the outer extremity in order to balance the system and to counter cable compression caused when the car travels up the cable. The system will be kept taught by the centrifugal effect of the planet's rotation. The SkyStation would be positioned on the cable at halfway (17 034km) with its central mass being above the level of areostationary orbit. Transit up the cable would be via a 'car' propelled via a passive Maglev drive system which would negate the need for the tether cable to be powered and therefore reduce the strain on and mass of the tether cable. The car itself will be of aerodynamic design similar in appearance to a high speed train and would travel at hypersonic speed (due to the 17 034km travel distance). It is envisaged that the transit time would be in the region of 3 hours taking into account acceleration and deceleration times giving an average speed of approximately 5 600km/h.

The car would be capable of transporting passengers and a cargo load of around 20 000kg.

- Transit between Equatorial Base Bravo and SkyStation
    - Cable compression avoided with counterweight at extremity at 34 000km
    - Central mass kept above level of areostationary orbit
    - Centrifugal force keeps tight and balances forces
  - Constructed from Fullerenes and carbon nanotubes and Diamond nanothreads
    - Fullerenes: carbon molecules in form of spires and tubes. Maximum strength = geodesic shell (60 carbon atoms – Buckmister Fuller Geodesic Domes)
      - Carbon nanotubes: 250 x stronger and lighter than titanium
      - Tether constructed of carbyne carbon allotrope and diamond nanothreads. These can be constructed by compressing benzene to 200 000 times atmospheric pressure and then when the compression is released slowly they reform into pyramid-like tetrahedrons which are similar in structure to diamonds (63 GPa)
- Car configuration
  - Aerodynamic tube – similar to HST
    - Passenger accommodation
    - Cargo compartment
      - Estimated passenger and cargo capacity of 20 000kg
  - Automatically operated
  - Stabilisation utilised to counter Coriolis effect
- Power plant
  - Passive Maglev System via conductive tether
  - Decelerates prior to entering atmosphere to avoid friction and need for heatshield

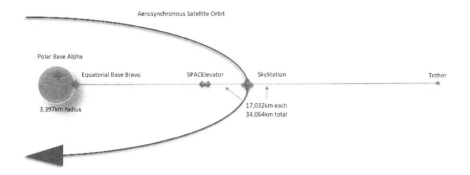

Figure 11. *Illustration of principal structural units*

Polar Base Alpha: *Principal base*
Equatorial Base Bravo: Planetary arrival port for astronauts
                                            including tourists) and cargo
                                        Logistics centre for distribution of
                                                supplies to Polar Base Alpha (8 000km
                                                distance – connected by Hyperloop) and
                                                exploratory camps
SkyStation: *Transit/transfer point from spaceships to*
                                                Space Elevator for astronauts and cargo

6 Technical Solutions

6.1 Oxygen Generation / Supply

Oxygen for the habitat will be supplied from the decomposition of the carbon dioxide within the Martian atmosphere. The system will compose of two elements, firstly the storage and utilisation of the atmospheric CO_2 to generate power from renewables and secondly the recapture of the exhaust gas from the power generation for low temperature plasma decomposition into oxygen and carbon elements.

There are several methods that can be used in order to convert CO_2 including catalytic conversion, photo-catalytic/electrochemical processes, enzymatic/biochemical processes and plasma processes. Each method provides differences in the products obtained

Use of low temperature plasma decomposition has benefits due to the reactor size in comparison with thermal plasma processes, along with higher thermodynamic efficiency and a near zero emission profile. Non-thermal plasma (NTP) utilises electrons to excite the molecular and atomic species and by doing so breaks up the chemical bonds without the need for high temperatures. A traditional thermal process would require high temperatures within the $1\ 600^{\circ}K - 2\ 000^{\circ}K$ ranges in order to dissociate the molecule.

NTP also gives high reaction rates and a steady rate of reaction is quickly achieved providing stability and reliability for the supply of oxygen. An added advantage of the NTP process is that due to the stable and steady characteristics, the system would facilitate rapid start-up and shutdown which would be advantageous when linking the process to renewable sources of energy such as wind and solar as described in Section 6.4.1 [12].

There were three types of NTP considered:
    i.   Glow discharge (30% conversion at an input of 7kV) [13]
    ii.   Radio frequency discharge (90% conversion with 100W input) [14]
    iii.   Dielectric barrier discharge (DBD Reactor approx. 30% conversion at a power density of $14.17\ W\ /\ cm^3$) [15]

As with all plasma systems, a balance has to be reached between the conversion of CO_2 and the energy required to obtain an efficient conversion rate. To solve the issue of producing an energy efficient system, which is made all the more necessary as the NTP system will be linked predominately to the renewable power sources, a modification to the standard plasma system would be required. This modification would be in the form of the addition of a catalyst into the plasma discharge. Research has shown that the incorporation of a catalyst can result in the increased conversion of CO_2 whilst being able to maintain a low energy consumption profile [16].

The type of NTP to be utilised within the colony is a DBD reactor with the incorporation of a catalyst into the plasma discharge which gives higher CO_2 conversion whilst maintaining thermodynamic and efficiency. This type of reactor also has the benefits that changing the catalyst will alter the chemistry and therefore the interactions between the plasma and the CO_2 producing more complex carbon-based products, such as formaldehyde, methanol, ethylene and C_4 hydrocarbons. It is envisaged that the use of such a system would give an added degree of flexibility and potential multiple production possibilities [17].

The reactor(s) which will be incorporated into the renewable power and atmospheric process will be a hybrid single stage plasma-catalytic dielectric discharge reactor.

In the single stage reactor the catalyst can be placed directly into the discharge in order to obtain increases in efficiency and production of other carbon products. It is possible to hydrogenate the CO_2 in plasma by utilising a reaction with hydrogen or water at atmospheric pressure within the plasma catalytic processor discharge without the requirement for the high temperatures and pressures that a conventional thermal catalytic process would require.

Methanation is shown in Equation 1 and the reverse water-gas shift reaction is shown in Equation 2.

Hydrogenated CO_2 $CO_2 + H_2 - CO + H_2O$ (eq.1)

Methanation of CO_2 $CO_2 + 4H_2 - CH_4 + 2H_2O$ (eq.2)

Hybrid Plasma-Catalytic Process

Below is a simple diagram of a hybrid plasma-catalytic reactor where the catalyst could be substituted in order to achieve production of the desired product. It is envisaged that the primary function would be the generation of oxygen for the LSS, however the system could be used in a parallel configuration with a manifold supplying the atmospheric CO_2 to processors producing different end products.

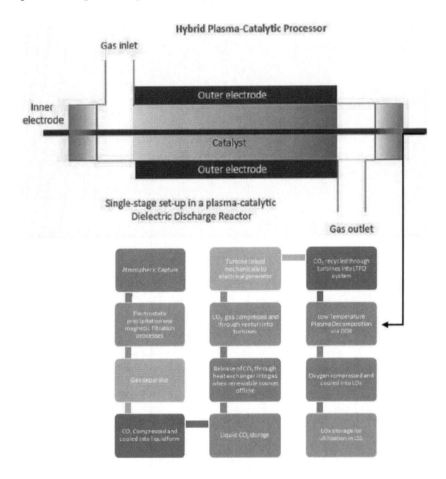

Figure 12. *Diagram of a Hybrid Plasma-catalytic Reactor*

6.2 Water Resources

Water will be sourced primarily from water ice available adjacent to the Polar Base Alpha at the northern polar region. The ice will be melted within a sealed chamber which has been drilled into the ice cap and then pressurised allowing a heating element to be introduced and the liquid water to be obtained as illustrated below:

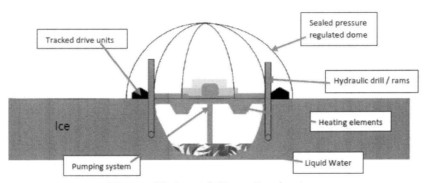

Figure 13. *Liquid Water Production*

As the liquid is potentially contaminated with fine dust material which has been trapped within the ice, a filtration system will be employed in order to remove particulates before the water is purified, boiled and condensed resulting in the production of distilled water for use within the colony.

This water will then become part of the colony's water management system where as much of it as possible will be recycled, minimising the need for extraction of natural resources to minimise the impact on the Martian environment.

A diagram below shows the basic configuration of the colonies water management system.

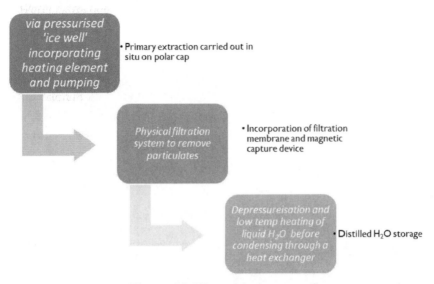

Figure 14. Water Management System

The water recycling components of the system ensure that the impact of the extraction from the environment are minimised. The system separates H₂O into two streams: one for hygiene and systems, and the other for potable H₂O.

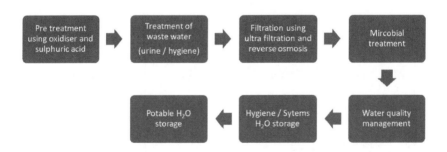

Figure 15. *Physio-chemical water recycling solution for Polar Base Alpha*

6.3 Food Production

Food production within the colony will consist of a multi system approach. This will provide variety in diet and at the same time limit the risks in the event of a single system failing. Key elements considered for food production requirements within the colony were:

- Selection of crops for production (including GM elements)
- Environmental – sunlight, radiation, pressure, gravity, temperature etc.
- Water delivery to root zone due to low gravity
- Recycling and the addition of nutrients/minerals from waste materials

These elements mean that the food production section of the colony, located at Polar Base Alpha would consist of the following systems:

- Plant growth via processed basalt Martian soil with added nutrients in specialised greenhouses
- Plant growth via aquaponic and integrated hydroponic (aeroponic) systems
- Artificially produced (lab based) products and nutritional supplements

6.3.1 Soil Based Food Production

The Martian soil (basalt based materials) will be the foundation of the soil based crop production. This will be extracted and processed by robotic agricultural units as described later in Section 9.

The Martian processed soil will then be enhanced by the incorporation of materials from a bio-reactor and combined with a nutrient rich solution derived from the urine processing system. This will also be supplemented by compost generated from vermiculture.

Bioreactor
This will comprise of the main reactor with the shredding and dehydration element which can recycle and transform waste food and inedible plant matter (including root and stems) into a rooting structure for the next growing cycle.

41

Hydration

The water and hydration will be via direct injection into the soil and rooting substrate in order to overcome issues associated with the low gravity environment. The water will be contained in collapsible reservoirs and the fluid flow will be controlled via pumps.

Pollination

Where pollination is required this will be done through Smart Mote drone technology (mechanical insects). These will be released in greenhouse environments as required and will be controlled via the central Agri-computer systems.

Sunlight

As the sunlight reaching the Martian surface is less than the sunlight which reaches the surface of Earth, supplemental light is required. It is envisaged that artificial solutions mimicking 'natural light' will be incorporated into the greenhouse structure. This will be vital due to the high latitude of Polar Base Alpha. The solar mirror could be deployed to assist with the sunlight levels, however the effects would be marginal due to the deflection angle.

The types of planting considered for the soil based food production are:

- Wheat (winter)
- Peanuts
- Sweet Potato
- White Potato
- Soybean

These were considered due to dietary requirements and also on preferred CO_2 concentration, resilience and lux requirements

6.3.2 Aquaponics (Aeroponics)

The aquaponics systems to be used at Polar Base Alpha combine conventional aquaculture with hydroponics to achieve a self-sufficient, symbiotic food production system. Aquaponic systems are low energy and generally low maintenance and can provide a wide variety of crop production with high crop yields. It is further enhanced by the use of Aeroponic techniques.

42

The aquaponics system requires live components to work successfully. The three live components used in the colony are:

- Fish
- Plants
- Bacteria

The selection and type of plants being grown depend on the type and density of the fish stocking being utilised. With this in mind a multi-bay system will be used in which the fish density will differ according to the crop being cultivated. Where fish density is greater more nutrients from effluent is available and therefore more is available to be recycled to the plant roots via the bacterial processing.

The types of crops produced in the multi-bay system are shown below

Low-Medium Fish Stock Density	
Green Leaf Vegetables	
Chinese Cabbage	Turnips
Cabbage	Parsnips
Beans	Cauliflower
Peas	Broccoli
Spinach	
Herbs	
Sage	Lemongrass
Coriander	Basil
Parsley	Chives
High Fish Stock Density	
Salad Plants	
Cucumber	Capsicum
Shallots	Red Salad Onions
Tomatoes	Snow Peas
Chilies	Strawberries

High nutrient requiring fruiting plants such as Melons can also be cultivated where fish stocking densities are increased further.

The Aquaponic system has the following parts:

These two parts are then broken down into several component parts or sub-systems which are responsible for the effective removal of solid wastes, for adding bases to neutralise acids and for maintaining water oxygenation.

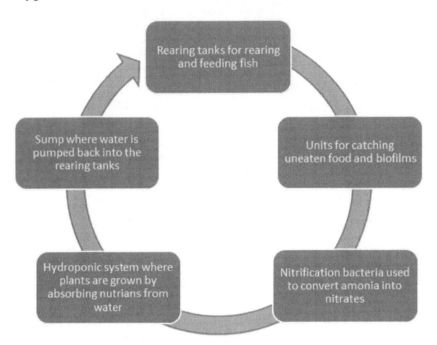

Figure 16.*Aquaculture system*

Rearing and Fish Stocking
The colony's systems utilise freshwater fish due to the non-saline nature of the water recycling processes and due to the fact that freshwater fish are able to tolerate crowded conditions. For the purposes of food production Terra Nova will use selective species which are non-toxic to

humans and can be harvested by colonists for consumption. The fish chosen for the aquaponics system are

- Channel Catfish
- Rainbow trout
- Perch
- Common Carp
- Arctic Char
- Bass (large mouthed / striped) [18].

Food for the fish will come from two sources:

- Growing of duckweed within the hydroponic system itself (thereby achieving a system close to self-sufficiency)
- Use of excess worms used within the vermiculture composting process as part of the larger food production system

This means that the fish diet is supplemented by using the prepared kitchen/food/ plant matter scraps generated by the colony through the composting process. The broken down compost from the vermiculture process will be used as nutrients for the soil based plant cultivation.

Settling Tanks

These will be incorporated into the fish tanks and will utilise locally sourced pumice material, perlite as well as synthetic granules to aid in the capture and extraction of material.

Bio-filtration Subsystem

Bacteria in the process is vital for the aerobic conversion of ammonia which is constantly released into the water by the excretion of the fish and as a product of their metabolism. As even small increases of ammonia damages the system, an efficient and effective filtration system is required. The bio-filtration system will include *Nitrosomonas* bacteria to convert the ammonia into safer nitrogenous compounds. The bacterial process nitrifies the water which raises the pH. As part of the filtration subsystem additives such as potassium hydroxide or calcium hydroxide can be added to balance and neutralise the waters pH.

6.3.3 Hydroponic (Aeroponic) Subsystem

Aeroponic techniques are utilised in order to reduce the use of water in the growing bags and also to increase the amount of air to the plants. As the microclimate within the bags can be finely controlled when using aeroponic techniques and as the plants are in suspended galleries, they are able to receive 100% of the available oxygen and CO_2 to the roots zone, stems and leaves. This has the advantage of accelerating biomass growth (quadrupling growth rates) and reducing rooting times.

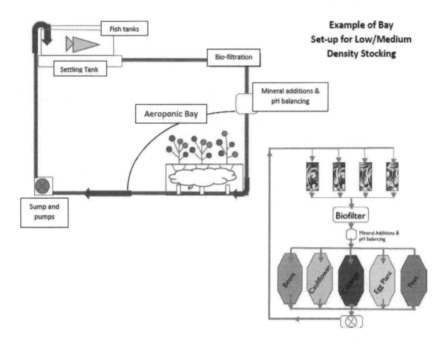

Figure 17. *Aeroponic System*

The aeroponically grown plants within the bays will have an 80% increase in dry weight biomass when compared to standard hydroponically grown. This is all achieved without the need for any chemical fertilisers or pesticides.

The aeroponic bays will be vertically stacked in racks to maximise the use of space and each rack will be fitted with specialised red and blue lighting. This lighting can be fine-tuned in order to ensure energy efficiency and control growth rates of the plants.

Aeroponics also have an added advantage as due to the low gravity conditions experienced on Mars the nutrient and water interface with the plants can be controlled more effectively when they are introduced into the system as an aerosol. This is further aided in the process by natural wicking and capillary forces.

Water Usage
The aquaponics system and subsystems will not discharge or exchange water and will recirculate and reuse water contained within the circuit. The only water which will be added will be to replace the losses encountered due to evaporation processes.

The water top-up will come from the waste water recycling processes within the colony. This predominately will be from the urine processing whereby the stream is split at stage 3 from the main process and the waste urine is converted into a nutrient laced solution which is inserted into the hydroponics subsystem [19].

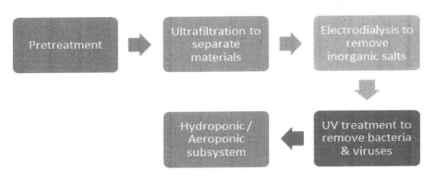

Figure 18. *Hydroponic System*

Any excess solution not required for the Hydroponic subsystem will be used in the soil based plant production process.

6.3.4 Laboratory Cultured Protein

As well as the food produced from the agri-greenhouses and aquaponics systems the colonists' diets will be supplemented by artificially grown laboratory protein and meat products. This will ensure a varied and sustainable diet can be maintained by them whilst on their two year rotation on Mars.

The types of cultured meat to be available will be chicken, pork turkey and beef due to their varied taste and nutrient profiles.

The laboratory facilities for the culturing and production of the products will be located at Polar Base Alpha within the agricultural section. The culturing of product will be achieved by importing frozen muscle specific stem cells from animals on Earth which would then be cultured into fully developed fibres within the laboratory, producing the food product for consumption by the colonists.

It is envisaged that these *in vitro* muscle protein production and tissue engineering processes will use mytoblast cells. These types of cells are ideal as they are able to proliferate at a rapid rate and at a sufficient stage of development to allow ease of growth into the desired product.

Once implanted into the process, these cells are then treated by adding a protein that promotes tissue growth before being transferred into a culture medium for growth.

As meat tissue is three-dimensional the cells are grown in an edible scaffold which stretches with the developing tissue replicating natural animal growth.

The process will be undertaken in laboratory conditions within specially designed bioreactors.

These cultured proteins can be harvested from rapidly self-proliferating, self-healing mytoblast cells. A single cellulite cell can undergo as many as 75 generations of division within a three month period. From this type of cultivation a single turkey cell could be turned into twenty trillion turkey nuggets! [20].

The culturing process has many beneficial aspects which are not found in live animal farming and meat production, including nutritional and health benefits.

As the biochemistry can be manipulated, the product can have its nutrient profile tweaked in order to adjust and tailor it to the requirements of the colonists' diets which will differ from that of the same person living on Earth. Tweaks could include adjusting the fat to protein ratios, with significant beneficial health outcomes.

As well as the adjustment of the nutrient profile the cultured meat product(s) will have not been subject to the medications administered to the animal which they are associated with, such as antibiotics, and therefore there will be no trace contaminants within the meat. They will also not have come into contact with any chemical contaminants which may be present in animal feed or in the environment. The cultured meat will also be free of any bacterial contaminants such as salmonella as the product will be grown in sterile laboratory conditions.

6.4 Power Generation

6.4.1 Solar / Wind / Renewables ISRU

It is envisaged that the main supply of renewable energy sources will come from the facilities at the Equatorial Base Bravo as the latitude of Polar Base Alpha does not lend itself to efficient production of power via solar or from wind generated sources. The Martian winter and the reduced sunlight arriving at the geographical North location means that solar energy would be confined to secondary provisions only for Polar Base Alpha.

The Polar Base Alpha would utilise 5kV units to supplement power to local environments such as the food production areas and greenhouses. These units would further be enhanced and supported by the use of transparent photo-voltaic salts embedded into the domes to capture solar energy and convert it into electrical power which can then be fed back into the main grid system.

Equatorial Base Bravo will utilise solar and wind power being captured in solar arrays and through the use of tethered wind turbines and fixed installations.

The solar array(s) will be aided through the concentration of solar energy from an orbital solar mirror with an inclination angle optimised to capture and focus the sunlight onto the receiving surfaces.

Solar Mirror

The solar mirror works by altering its shape and angle in orbit and deflecting the solar energy onto the desired position on the surface.

The solar mirror will be positioned in an areostationary orbit above Equatorial Base Bravo in order to maximise the level of solar energy reaching the photovoltaic solar arrays positioned at the equator.

The orbital mirror will not only intensify the sun's energy reaching the arrays but also prolong the effective hours of daylight and therefore the effective generating time of the arrays.

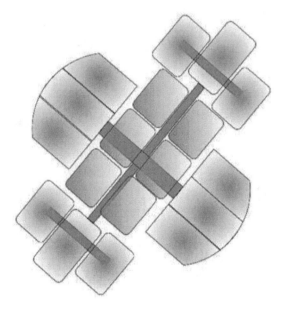

Figure 19. *Solar Mirror*

Tethered / Vertical wind turbines

These will generate power from utilisation of higher atmospheric wind conditions through the use of floating tethered wind turbines. The part ducted fans would be surrounded by ultra-light weight inflatable structures which are filled with hydrogen gas and then secured into the ground via tethers. The tethers could either be fixed or be positioned into a turntable mechanism in order to allow for any change of wind direction.

The power from these installations could be transmitted via microwave power beaming or through hard wire cabling.

In addition, conventional vertical bladed turbines at ground / low level would be employed to generate power. This would be fed into Equatorial Base Bravo's Central Power Supply and Distribution System (CPSDS).

The CPSDS would manage and distribute power and deal with the switching and transforming of power where required. It would also deal with the transmission of power through the microwave distribution network to Polar Base Alpha and for the purposes of supporting the robotic operations.

As the consistency of ISRU renewables is not guaranteed, the system will be integrated into the Oxygen and Carbon capture process to allow for the generation of electrical power when either wind or solar energy is not available.

6.4.2 Fission Reactor Option

Utilisation of Uranium 235 via a PWR (Fission)

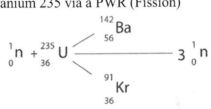

$$^{1}_{0}n + {}^{235}_{36}U \rightarrow {}^{142}_{56}Ba + {}^{91}_{36}Kr + 3\,{}^{1}_{0}n$$

A standard nuclear fission reactor could be employed where Uranium 235 is bombarded with neutrons forcing the atoms to split. This creates two smaller atoms of Krypton and Barium along with large amounts of energy plus 3 spare electrons, which in turn create a nuclear chain reaction.

Pressurised Water Reactor

In order to utilise the in-situ resources as far as possible a pressurised water reactor will be constructed and sited within the engineering quarter of the Polar Base Alpha. It is envisaged that this will be positioned underground along with the associated heat exchangers and turbines as well as the condensing and recapturing system.

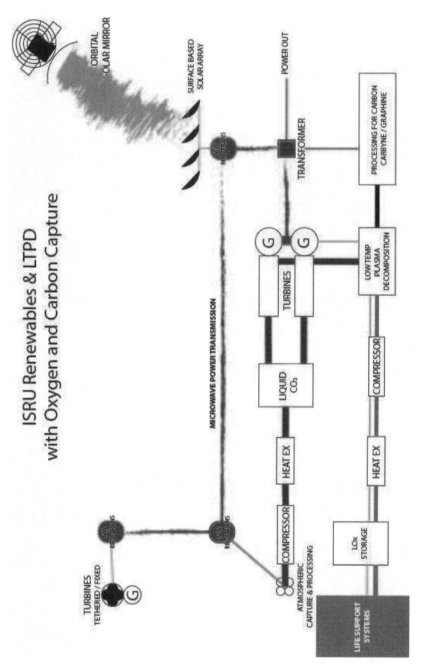

Figure 20: *ISRU renewables and LTPD with oxygen and carbon capture*

The reactor pressure vessel and control rods would be constructed within a containment structure built from locally sourced materials. This material could consist of regolith, dust extraction slurry and rocks which are then formed into blocks via microwave technology.

The water for both the open and closed loops would be supplied from the H_2O processing plant with the warm condenser water being recycled for other purposes within the colony.

Power from the reactor would be supplied via steam from the heat exchangers being passed through multi-stage turbines linked to generation equipment. The power would then be transmitted to the Central Power Supply and Distribution System (CPSDS).

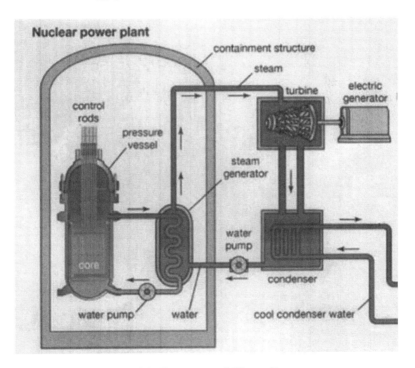

Figure 21. *Pressurised Water Reactor*

The drawbacks with the utilisation of such a power plant are that fission involving Uranium 235, Plutonium 239 and even Thorium 232 produce high levels of radioactivity hazards both in the production of the power through the exothermic reactions contained within the reactor core and through the production of highly radioactive waste products.

The shielding and containment required to house the reactor and the ancillary systems requires a high level of resource and would also require supply of fuel / refuel and disposal of waste for reprocessing via shuttle to Earth.

This was considered as a secondary power supply option but was not selected as the preferred option due to due the numerous drawbacks.

6.4.3 Fusion Reactor Options

Nuclear fusion takes place when light atomic nuclei are propelled at sufficient velocities into one another so that when they collide they combine overcoming the electrostatic force repulsion, to form a heavier atomic nucleus and releasing a large amount of energy. In order for the fusion reactions to take place, there is the need to have a device which can achieve sufficient rate of kinetic energy and confinement.

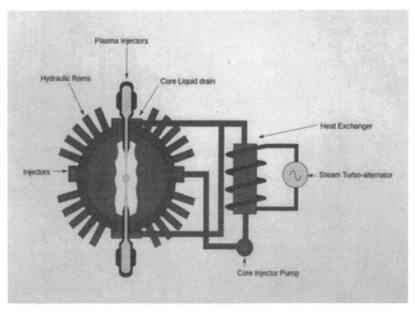

Figure 22. *Diagrammatic description of a fusion reactor with a steam powered turbo alternator used to generate electrical power.*

Nuclear fusion has significant advantages over the fission model in terms of facilities required for generating the power and thermodynamic efficiency and through the by-products and waste produced. Conventional neutronic nuclear fusion reactors utilising Deuterium and Tritium are relatively benign in regards to having no long term waste issues associated with them and they produce high yields of energy by mass. There is also the added safety considerations that the reaction has high energy density but cannot 'blow-up or melt down' and it requires much less footprint and is significantly more compact than a fission type reactor.

Described below is the Tokamak device which was considered as an option utilising Deuterium and Tritium which gives off around 80% of its energy in the form of fast neutrons, making the apparatus have a drawback of being relatively radioactive in the short term. In both cases the waste material is negligible in comparison with Uranium and any stray neutrons have a half-life of between 50-100 years. Both of these isotopes are easily obtainable as Deuterium can be obtained from water and Tritium can be obtained from Lithium.

To effect fusion using a Tokamak reactor, hydrogen is heated to more than 100 million degrees Kelvin to obtain a plasma which is then contained within a magnetic field. The plasma is contained within Tokamak device which uses high powered magnetic confinement to control the fusion reaction and produce the thermonuclear reaction. This reaction produces vast amounts of exothermic energy which is harnessed and converted into electrical power [21].

The Tokamak design considered has a drawback that the energy conversion rate is around the 30 percentile mark and it uses magnetic compression which is also inefficient.

This type of design was considered as a tertiary option.

Magnetic and Electrostatic Nuclear Fusion Reactor (CrossFire Fusion Reactor)
It is envisaged that the main power source and supply for the Mars bases would be via a CrossFire nuclear fusion reactor. This type of reactor is a development on the standard design of fusion reactor in that it is able to

produce usable fusion energy at significant rates and thus meet the power requirements of the facilities.

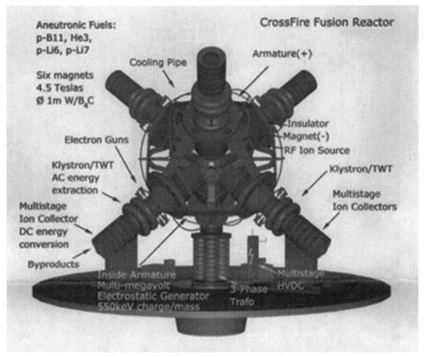

Figure 23. *CrossFire Fusion Reactor*

The design can take full advantage of the electrostatic acceleration and has the ability to use multiphase electrical currents flowing through concentric coils. This allows the reactor to both accelerate and confine the plasma in order to make it more efficient and effective for producing enormous amounts of electrical energy from clean safe and environmentally low impact aneutronic fuels.

Aneutronic fusion is clean and safe and only a minimum of radiation shielding is required and therefore it is an excellent option for power generation for the Martian bases. As most of the energy produced by aneutronic fusion is in the form of charged particles instead of neutrons, which can be converted directly into electricity by making them work against the electric / magnetic fields, this has the possibility of exceeding 90% efficiency [22].

The CrossFire reactor utilises six super conducting magnets to form a magnetic cusp region where positive ions can be injected. In this cusp region a negative voltage is applied and a positive voltage is applied at the opposite end of the magnet. This causes the ions to be electrostatically accelerated towards the negative potential passing through the magnetic cusp into the interior of the chamber where they are confined by magnetic and electric fields. The ion injection is continuous surrounding the magnetic cusp region which allows for a 3D injection into the chamber [23]

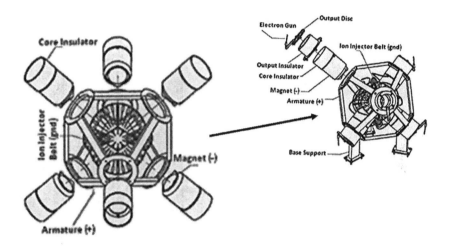

Figure 24. *Exploded view of the CrossFire reactor core chamber [23]*

The positive voltage is then controlled to confine only reactants within the chamber thus allowing the charged products from the fusion reactions to escape and be directed for electrical production.

The power generation will be via two forms of energy capture
Energy conversion from positive ions into electricity by creating a positive electric field to slow down the ions converting their kinetic energy to potential energy and an electron gun to neutralise them. The electron gun extracts electrons versus the positive current.

Cooling medium being circulated around the reactor would be directed through heat exchangers and the resultant heat energy released used to produce HT steam. This steam would then pass through a series of staged turbines to convert the mechanical energy into electrical power.

The table below shows aneutronic reactions which are of notable interest due to production of charge particles in the primary reactions that can be directly convertible into electrical energy [24].

Reactants	Products	Energy Density
$^1H + 2\ ^6Li$	$\rightarrow ^4He = (^3He + ^6Li)$ $\rightarrow 3\ ^4He + ^1H + 20.9$ MeV	153 TJ/kg 42 GWh/kg
$^1H + ^7Li$	$\rightarrow 2\ ^4He + 17.2$ MeV	204 TJ/kg 56 GWh/kg
$^1H + ^9Be$	$\rightarrow ^4He + ^6Li + 2.1$ MeV	22TJ/kg 6 GWh/kg
$^3He + ^3He$	$\rightarrow ^4He + 2\ ^1H + 12.9$ MeV	205 TJ/kg 57 GWh/kg
$^1H + ^{11}B$	$\rightarrow 3\ ^4He + 8.7$ MeV	66 TJ/kg 18 GWh/kg

Calculations on potential energy output(s)

Electronvolt $\rightarrow$ Joule : $1eV = 1.60218 \times 10^{-19}$ J

Electronvolt $\rightarrow$ Temp : $1eV = 11\ 604.505$ K / $11\ 331.355\ ^0C$

Electronvolt $\rightarrow$ mass : $1eV = 1.782662 \times 10^{-36}$ kg

Electronvolt $\rightarrow$ mass : $1MeV = 1.782662 \times 10^{-30}$ kg

Particle	Mass	Charge
Proton Mass	1.67262×10^{-27} kg	$+ 1.60218 \times 10^{-19}$
Neutron Mass	1.67493×10^{-27} kg	0
Electron Mass	0.00091×10^{-27} kg	$- 1.60218 \times 10^{-19}$

Example of potential energy output using Pentaborane (B_5H_9)

Pentaborane mass: $5 \times 18.41723\times10^{-27} + 9 \times 1.67353\times10^{-27} = 107.14792\times10^{-27}$kg

Specific Energy of Pentaborane:
$5 \times (8.68\text{MeV} - 123\text{keV}) / (107.14792\times10^{-27}\text{kg}) = \underline{3.99308\times10^{32}}$ eV/kg

Specific Energy (J/kg): $3.99308\times10^{32} \times 1.60218\times10^{-19} = \underline{63.97633\times10^{12}}$ J/kg

Specific Energy (GWh/kg): $63.97633\times10^{12} / (3.6\times10^{6}) = 17.77120\times10^{6}$ kWh/kg $= \underline{17.77120 \text{ GWh/kg}}$

Extracting 3 electrons from Pentaborane to produce positive ions:
$107.14792\times10^{-27} -3\times0.00091\times10^{-27} = \underline{107.14519\times10^{-27} \text{ kg}}$

Charge to mass ratio of Pentaborane (C/kg) after extracting 3 electrons:
$3\times1.60218\times10^{-19} / 107.14519\times10^{-27} = \underline{+4.48600\times10^{6} \text{ C/kg}}$

The specific energy and charge to mass ratio are essential parameters to define the magnetic flux and electric voltages.

Using the specific energy to find the velocity of products from the nuclear reaction:
$E=\frac{1}{2}mv^2 \rightarrow v = (E/m) \times 2)^{0.5} \rightarrow v = ((63.97633\times10^{12}) \times 2)^{0.5} \rightarrow v = \underline{11.31162\times10^{6} \text{ m/s}}$

Calculating the specific impulse:
$11.31162\times10^{6} / 9.80665 = 1.15346\times10^{6}$

(Defining the magnet bore of about 900mm - or .9m with internal radius of .45m)

Using the charge to mass ratio calculation to find the magnetic flux:
$R=mv/qB \rightarrow r = (v/B) \times (m/q) \rightarrow r = (v/B / (q/m) \rightarrow B=v / (r \times (q/m))$

$B= 11.31162\times10^{6} / (.45 \times 4.48600\times10^{6}) = \underline{5.60341 \text{ Teslas}}$

Therefore a superconducting magnet of at least six Teslas (or higher) and with a bore of 900mm is sufficient to radially confine the plasma of both the reactants and products.

Calculation of a negative voltage for electrostatic acceleration of the positive ions to gain enough kinetic energy to reach at least 123keV, hence 550KeV should be enough:

$E = qx\ V \rightarrow V = E/q \rightarrow V = (E/m) / (q/m) \rightarrow$

$V = ((5 \times 550KeV \times 1.60218x10^{-19}) / 107.14519x10^{-27}) / 4.48600x10^6 =$ 916.66667x10³ Volts

Calculation of Temperature:
550x103 x (11604.505 K -273.15) = 6.23224x10⁹ ⁰C

A negative voltage of -920 kV is sufficient for the positive ions to gain the required kinetic energy which would be equivalent to 6.2 billion ⁰C (approx.) Calculation of a positive voltage in order to trap (longitudinally) the reactants and allowing the charged products to escape is a kinetic choice between the reactants 550KeV and products 8.68MeV and could be something around 1.5MeV.

$E = q \times V \rightarrow V = E/q \rightarrow V = (E/m) / (q/m) \rightarrow$

$V = ((5 \times 1.5\ MeV \times 1.620218x10^{-19}) / 107.14519x10^{-27}) / 4.48600x10^6$ = 25 000x10³ Volts

$V = 2500x10^3 - 920kV = 1\ 580x10^3$ Volts

A positive voltage of 1 580kV is enough to trap the reactants whilst allowing the products to escape.

Calculating the fuel consumption:
The consumption of a fusion reactor at a power of 500MWatts using Pentaborane which has a specific energy of 63.97633x10¹² J/kg is...

500MW = 500x10⁶ J/second →
      500x10⁶ J/second / 6 397 633x10¹² J/kg = 7.81539x10⁻⁶ kg/s

Therefore a fuel consumption of 7.82 milligrams per second is required for the production of 500MW

It should be noted that the calculations above relate to the direct energy produced by the products of the reaction and do not take into account the secondary energy production system of utilising the thermodynamic properties of using the coolant to produce high temperature steam via utilisation of heat exchangers and staged pressure steam turbines linked to electrical generators. As an approximation (based on current fission linked generation systems) the power generated from the secondary system(s) associated with the fusion option could yield another 1 000MW.

6.4.4   Radioisotope Power Systems (RPS)

Static & Dynamic RPS
It is envisaged that there will be a variety of applications where RPS systems will be utilised on the Mars bases. These will in the main be used to power robotic devices, including the nanotechnology as described in section 9.

Historically the use of static thermoelectric designs have been used in spacecraft and these have proved to be highly reliable, however these generally offer low energy conversion efficiency. It is due to this fact that the use of this type of RPS would be restricted to use in the types of applications listed below:
o   Nano-climites
o   Nano-drones
o   MEMS (micro-electrical mechanical systems)
o   Atmospheric and environmental monitoring equipment
o   Small android applications

Generators based on dynamic power conversion is another option for use in applications both as a back-up to the Beamed Microwave Power Distribution Grid and as a primary power source. As dynamic radioisotope generation carries a higher risk of failure due to the number of active coolers associated with their design, these will be limited to certain types of applications falling within the immediate compass of the surface base facilities. This will ensure that any issues over reliability is mitigated.

Multi-Mission Radioisotope Thermoelectric Generators (MMRTG)
As these systems have individual thermoelectric conversion elements numbering in the hundreds and potentially thousands they have and

excellent redundancy profile. The loss of a number of elements would have limited operational impact. These systems will be used in various applications and particularly ones that require higher power outputs. The MMRTG system will utilise Plutonium oxide as the fuel as power. Applications for MMRTG are considered below:

- o Poly-pedal exploration and analysis devices / robotics
- o Maintenance and Facilities Management devices / robotics
- o Agricultural and cultivation robotic units

We envisage that enhanced Multi-Mission Radioisotope Thermoelectric Generators (eMMRTG) based upon skutterudite compounds will be used in specific devices as these are more efficient and have extended life cycles [25]. Skutterudites have many heavy atoms and complex structures and are able to conduct electricity like a metal whilst at the same time insulate against heat in the same way as glass does. Due to these properties they are able to generate sizable electrical voltages and supply consistent and high energy electrical power both in medium and high temperature configurations. The eMMRTG systems would be utilised in devices and units such as:

- o Water resource / ice excavation
- o Use within the ISRU and production processes
- o Bi-pedal and polypedal units which require high power outputs

Radioisotope Thermophotovoltaic (RTPV) Systems
We envisage that there will be a need to use RTPV power generation systems in various orbital facilities and applications associated with the Mars project these will include:

- o SkyStation
- o Solar Mirror
- o Communications & GPS satellite constellation
- o Orbital MPDS units

As the solar flux around the Martian orbital path is around 30% that of Earth the orbital systems will require a reliable and consistent sources of power which is not reliant on solar energy. The RTPV system uses infrared radiation from a radioisotope heat source in order to generate electrical power through photovoltaic cells which can then be utilised by the on-board spacecraft systems. The proposed RTPV will be composed of 3 main elements:

- o Radioisotope Heat Source
- o Thermovoltaic Converter
- o Heat Regulation / Rejection System

The preferred Radioisotope heat source will be Plutonium Oxide (PuO_2) and would be housed within a standard design General Purpose Heat Source module (GPHS). This would be positioned within the spacecraft application in a completely sealed canister which in turn would be surrounded by photoelectric cells. These cells could be of a standard design or be further enhanced by the use of Phototonic Crystal Spectrum Control (PCSC) to realise a higher efficiency [26]. With the use of PCSC the emissions can be selectively controlled to match the TPV cells and therefore gain a higher efficiency.

This is illustrated in the Figure 25.

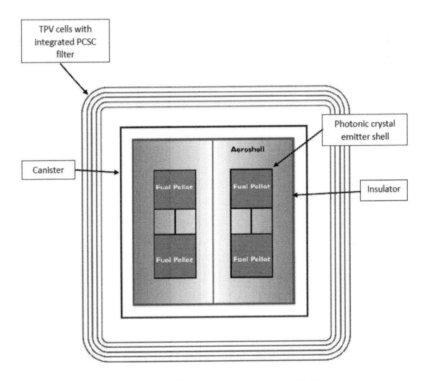

Figure 25. *Thermophotovoltaic Cell*

The system will utilise a minimum of 64 TPV cells per converter face which surround the GPHS in order to produce electrical power. Due to the generation of heat from the decay of the PuO_2 the system must have an integrated heat regulation and rejection element. This element is comprised of carbon-bonded carbon fibre fin structures with an aluminium honeycomb core. These fins are a vital component in achieving system efficiency, as RTPV performance is directly linked to heat rejection temperatures.

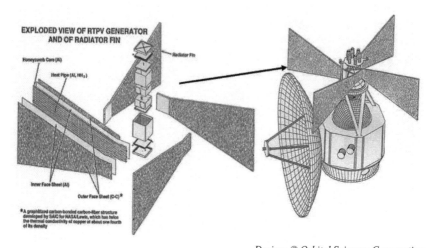

Design: © Orbital Sciences Corporation

Figure 26. *Radioisotope Thermophotovoltaic (RTPV) System.*

This type of assembly could be installed in a variety of spacecraft solutions as illustrated in this basic representation of the Pluto Fast Flyby spacecraft.

It is envisaged that a power of approximately 1.6 Watts(e) can be achieved per cm². This would be achieved with an operating cell temperature of around 1250°C. It should be noted that higher source temperatures and lower cell temperatures will increase power output [27].

6.4.5 Power Distribution

It is envisaged that power distribution will be carried out via 'hard wired' systems and via a beamed Microwave Power Distribution (MPD) network. The main colony's power will be supplied via either the fission or fusion reactor and this will be transmitted to the Central Power Supply and Distribution System (CPSDS) at 440V three phase AC. The CPSDS will house the step-down and demand control facilities and will distribute the power to the various quadrants of the colony.

All the hard wired infrastructure will have extra precautionary insulation particularly for external cabling and infrastructure due to the nature of the Martian atmosphere and the potential for arcing.

The total power demanded by the colony is expected to be met by the output of the reactor plus a 25% margin allowable for unforeseen and exceptional circumstances. The power will be supplemented locally by any solar energy captured power through the photovoltaic systems within the domes and through the food growing facilities.

Power transmission via hard wired cabling needs to be kept to a minimum due to the fact that cables are expensive both in terms of weight as well as being resource and material heavy. This is magnified by the potential extra insulation required to cope with the Martian environment. In order to minimise this and to provide power 'on demand' to remote locations and to high power usage devices and facilities a Microwave Power Distribution Network (MPDN) is to be established. This will allow electrical power to be beamed over significant distances or into orbit for redistribution.

It is via this method that both Polar Base Alpha and Equatorial Base Bravo can be linked together in a power grid despite the 5300km distance separating them. The MPDN will comprise of transmitters and receivers (rectenna) which can be either be in fixed locations on the planet's surface or on robotic devices.

Where transmitters and rectenna are in fixed locations they can be linked on the surface to form nodes in a relay transmission network, much like a conventional hard-wired grid with the advantage that the transmission distance can be significant between nodes. The relay transmitter and rectenna positions would be dependent on terrain and would require to be in line of sight (see example below).

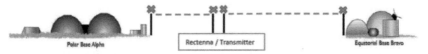

Figure 27. *Line-of-Sight microwave power transmission*

When power is required in a remote location for specific operations or to supply a mobile habitat then power can be beamed from the surface to an orbital platform. This platform can then radiate the power beam via a phased array to the facility or device requiring it. Where the device is out of range of the orbital array (past line of sight) the beam can be relayed via another orbital platform in order to reach the desired location (see example below).

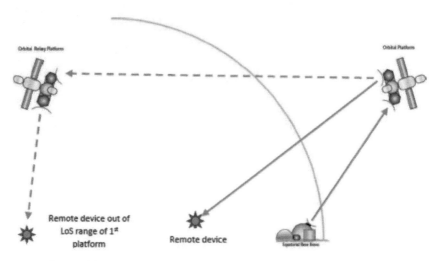

Figure 28. *Over-the-Horizon microwave power transmission*

As the maximum permissible human exposure to microwave radiation is ~25W/m², safety precautions would be instigated to ensure that the main transmission / rectenna do not pose a risk to the main habitats or to persons who were engaged in EVA.

7 Satellite Constellation

The colonisation of Mars will require the establishment of a significant communications, data and navigation network. Given the distances between Polar Base Alpha, Equatorial Base Bravo and the SkyStation together with any remote exploratory or mining expeditions, it would not be feasible to provide this through an areostationary satellite which would be unable to reach high latitudes. It is therefore necessary to use non-areostationary satellites in either elliptical or circular orbits. Areostationary orbit is the Martian equivalent of Earth's geostationary orbit, and is a circular equatorial orbit.

In addition, a deep-space relay will be required to provide continuous communication with Earth during conjunction when a line-of-sight Mars-Earth link is not possible.

Satellites will be assembled at Polar Base Alpha and launched through the SkyStation.

7.1 Areostationary satellites

- Line of sight communication with earth
- Deep space relay
  - Enables communication with earth, regardless of relative positions of planets
- Areosynchronous
  - Areostationary orbit = circular equatorial orbit @ 17,031.9 km provides coverage between 60° to 80.4° latitude depending on minimum elevation constraint.

7.2 Low Mars Orbit (LMO) micro satellites for relay of digital communications

- Relay satellite on inclined elliptical orbit
  - Enables global communications coverage with base station
    - Super-data rate (SDR) – 100 Mbps
    - Vox
    - Data

- UAV/RCV telemetry
- Relay communications to earth through stationary satellites
- Global Positioning System
- Reconnaissance/mapping

7.3 Communications

- Requirements
  - o 4 Satellites in circular orbit
  - o Altitude 4,800km
    - HDR-S = 1m antenna
    - LDR-S = 1.4m antenna
    - ISLs enabling connectivity
  - o High data rate link between Mars surface and orbit (HDR-S) – includes relay via aerosynch satellites for terrestrial elements outside line of sight (LoS)
  - o Low data rate link between Mars surface and orbit (LDR-S) – includes relay via aerosynch satellites for terrestrial elements outside line of sight (LoS)
  - o High data rate link between Mars terrestrial elements (HDR-M)
  - o Low data rate link between Mars terrestrial elements (LDR-M)
  - o Super-high datalink rate via wireless (short range) (SDR-W)
  - o Launch through SkyStation to minimise energy requirements

- Required for:
  - o Earth
  - o Remote static bases (exploration and exploitation)
  - o Manned SRDs
  - o Explorer vehicles
  - o Mars EVA astronauts
  - o Supply ships from earth (SkyStation and backup)
  - o Telemetry with UAVs, RCVs, robots

7.4 Navigation system (GPS)

- Requirements
  - 4 Satellites
  - 1 Sol HEO
  - Inclination angle $i=63.45°$ and eccentricity $e=0.71$

- Required for:
  - Mapping
  - Exploration
  - Control of robots and VAVs
  - Navigation of manned vehicles and aircraft

8 Transportation

Air transport will use SRDs of various sizes ranging from passenger and cargo transports down to UAVs utilised for mapping or delivery or small loads.

Land transportation will use TRVs: consisting of remotely controlled hovercraft for higher-speed, high-manoeuvrability missions; remotely controlled tracked vehicles with easy-change operational equipment tailored to requirements; and 'skidoo' type vehicles for use by a single astronaut.

All transportation equipment used on Mars will be remotely controlled from Polar Base Alpha through high speed datalinks using VR immersive systems.

A Hyperloop transport system will be used as the main connection between Polar Base Alpha and Equatorial Base Bravo.

8.1 Hyperloop Transport System

It is envisaged that a fast high speed transport system for the carriage of goods, materials and passengers will be required between the Polar Base Alpha at the latitudes of 80^0 and Equatorial Base Bravo. The distance between the locations is approximately 5300km therefore a fast reliable and efficient transport system is essential. The design of the system is based on the Very High Speed Transport System (VHSTS) which was proposed as early as 1972 and combines a combination of a magnetic levitation 'train' and a low pressure transport tube.

As the Martian atmosphere is already a low pressure environment (0.6 – 0.8 kPa) the need to have a complicated vacuum sealed and maintained tube system over such a long distance is overcome.

It is anticipated that the system would be enclosed to protect the users and components from both atmospheric conditions and from potential radiation concerns. Enclosure would be achieved by constructing an over ground tunnel built from in situ resource materials and utilising 3D printing and microwave construction techniques to form a hardened silica/iron tunnel sleeve. The sleeve will then be sub-compartmentalised to form two tracks (Northbound and Southbound) with a service tunnel in the centre for use in maintenance operations and for emergencies. Power for the system will be supplemented by the use of photovoltaic cells positioned along the length of the tunnel which would be maintained to achieve optimum efficiency by smart mote robotic units.

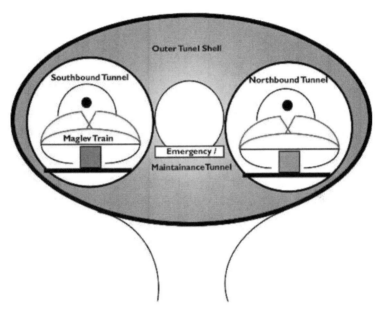

Figure 29. *Hyperloop System*

The system will utilise passive magnetic levitation, whereby the magnets are on the train carriage thus eliminating the need to power the tracks or the need for large scale copper coiling. As the Martian atmosphere cannot sustain human life the trains themselves will be fully pressurised and an airlock docking system will allow passengers and cargo to be transferred from the train to the terminals at either end of the system.

The maximum speed for the system will be approximately 1200km/h making the journey times from Polar Base Alpha to Equatorial Base Bravo at around 5 hours, after taking into account the need to maintain comfortable G Forces during the acceleration and deceleration stages of the journeys.

8.2 Air transportation

Semi-rigid Dirigibles (SRD)

Figure 30. *An example of a Semi-Rigid Dirigible*

- **Passenger and Cargo Carrier (PCC)**
  o VR remotely piloted (base station)
  o 100m long x 50m wide
    ▪ Tedlar/Mylar/fabric   sandwich   envelope
      construction
  o 60% aerostatic generation of lift; 40% aerodynamic
  o Electric or hydrogen fuel cell power contra-propfan
    ▪ Engines rotate on horizontal and vertical planes to
      give direction and thrust including hover
    ▪ Speed 350 kph
  o Mass of aircraft supported from strengthened beam built into
    hull during manufacture.
    ▪ Multiple attachment points on each part of structure
      allows load to be evenly distributed.
  o Pax seating and payload bay, attached to payload beam
    • Carry humans, robots, vehicles and other equipment
      o Tourism
    • Carry containerised/vehicle borne equipment

Unmanned Aerial Vehicle (UAV)

Figure 31. *Unmanned Aerial Vehicle*

- **Unmanned Aerial Vehicle (UAV)**
  - VR remotely piloted (base station)
  - Same construction and design principles as PCC
    - Speed 200 kph
  - Sensor array
    - Ground penetrating radar
    - LiDAR
    - Magnetic anomaly detector
    - Electro-optical/infrared and ultraviolet (EO/IRUV) sensors in chin mounted turret, enabling 360° observation
    - Forward looking infrared (FLIR) sensors with coaxially integrated laser pointer/ target designator
  - Communications
    - Relay real-time data back to base station

8.3 Terrestrial Rover Vehicles (TRV / MEV / RCV / SPV)

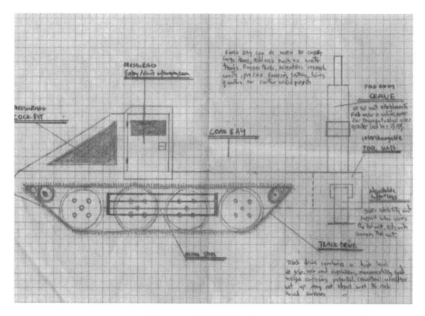

Figure 32. *TRV Concept*

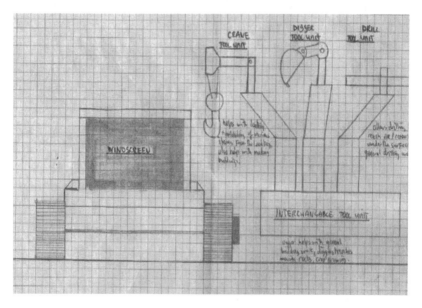

Figure 33. *TRV Tools*

Core design features/components
- o Wheels (solid or pneumatic tyres)
  - ▪ Medium transit speed
- o Tracked (full or semi)
  - ▪ Greater mobility/accessibility
  - ▪ Greater loads
  - ▪ Slower transit speeds
- o Hovercraft
  - ▪ Smaller, limited payload
  - ▪ Highest transit speed, but uses more energy/fuel
- o Power Plant
  - ▪ Gas turbine
  - ▪ Hydrogen fuel cell
  - ▪ Beamed power (MPD) Battery

- Manned Explorer vehicles (MEV)
  - o Driven by human
  - o Requires Life Support Systems
    - ▪ O_2 generator/CO_2 scrubber
    - ▪ Environmental control
      - Pressurisation
      - Air conditioning
        - o Dust prevention and removal
          - ▪ Magnetic filtration
        - o Temperature
  - o Airlock for EVA

- VR Remotely controlled vehicle (RCV)
  - o Used to transport robots and equipment

- Special purpose vehicles (SPV)
  - o Standard body, interchangeable components
    - ▪ Grader/bulldozer
    - ▪ Container for accumulating/carrying dust + earth for use in ceramic production etc. (road sweeper?)
      - Drilling
      - Extraction

- Scientific/Testing (laboratory)
- Medical/medevac
  - Air Transportable Surgical Unit (ATSU)
  - Hyperbaric chamber
- Living quarters/habitation – avoids need to return to base station or to set up long term base

Mars Chopper bike

Bike powered by battery electric motors capable of road / sandy conditions using a ski style steering. Deployable wheels lift the ski for use on harder surfaces when higher speeds can be achieved.

An extended version can be used as a medivac vehicle to transport persons in need of medical assistance. The load / cargo area can also be interchanged with a rear seat capable of carrying two people.

The vehicle has a quick charge time and long endurance giving greater flexibility meaning not having to use larger vehicles to conserve energy utilisation.

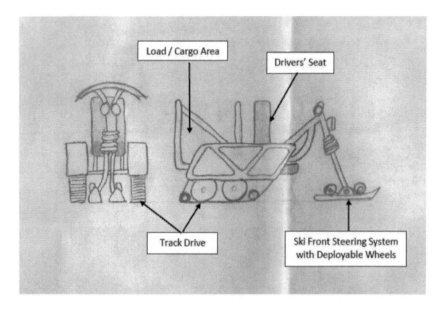

Figure 34. *Mars Chopper Bike*

- Running gear
  - Twin track on rear
    - Steering brake on track to assist turning circle
  - Ski steering with deployable wheels for hard/rocky terrain
- Coil spring front suspension
- Battery powered, utilising MPD grid
- Load bay can be swapped out for double rear-facing seat
- More manoeuvrable than larger vehicles – can be used for scouting

8.5 Mars Railer

The Mars Railer can combine tourist trips with research for fast efficient and safe travel. It will use pre-laid rails on designated routes. Although fitted with driver controls it can be easily automated for transporting goods and passengers. For both tourist and research trips several bikes or other transport systems can be loaded onto the load luger waggon to allow trips out with the rail network. Electric and hydrogen fuel cell hybrid set-up with solar panels provide power.

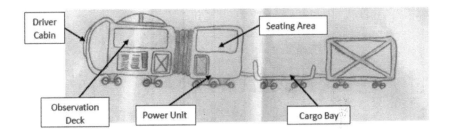

Figure 35. *Mars Railer*

- Slower although lower cost than Hyperloop
  - Automated
- Combined passenger and freight
  - Used to transport Mars Chopper bikes and other vehicles
- Automated system
- Powered by MPD system

8.6 Mobile Telescopic Habitat

The mobile telescopic habitat is a self-contained unit designed to be utilised for the short to medium lengths of time when away from the main bases. It has integrated LSS, accommodation as well as specific elements for exploration and research tasks.

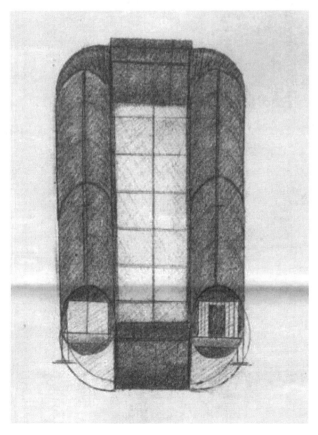

Figure 36. *Sketch of Telescopic Habitat*

The unit itself sits within a cylindrical chassis, which is loaded onto the back of a tracked vehicle and transported to the required location.

Once in situ the habitat telescoped outwards and around to form a 'U' shape. This is then anchored into the ground by tethers. The central area is covered in an inflatable transparent pressurised dome so that the inhabitants can utilise the space between the two legs of the habitat. The dome is impregnated with photo-voltaic salts to convert solar radiation to electrical power. Power can be provided by extending solar panels and via an on-board hydrogen system, however beamed microwave power can be utilised via an integrated power rectenna.

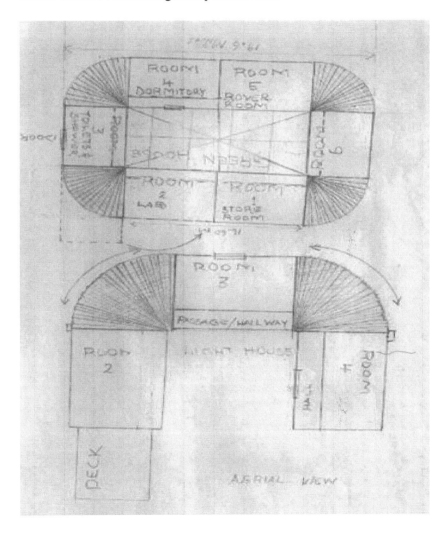

Figure 37. *Plan View of Telescopic Habitat*

Figure 38. *Habitat on Tracked Vehicle*

9 Robotics

9.1 Robot types

Various types and sizes of robots would be deployed to carry out tasks which would otherwise be too difficult, time consuming, labour intensive or have a high risk profile which would preclude a human operative from completing the task.

It is envisaged that Robotics will be deployed to assist in the running and maintenance of both Polar Base Alpha and Equatorial Base Bravo as well as in the production of products and materials. Specialised units will also aid in sensing and exploration as well as agriculture.

The size and design of the robotic unit will depend on the task it is applied to and these will generally fall into the categories described below:

Nanobots: Using nano technology at scales of a billionth of a metre or smaller. These devices could be deployed to administer medical functions or to process soil or waste materials. Nano technology is in essence molecular motors which can be autonomous and have single or multiple functions.

Smart Motes: These are robotic devices around the size of a grain of sand and can be deployed for monitoring of atmospheric conditions and to collect data. Both Nanobots and Smart Motes can be used in a 'swarm' capacity and co-ordinate responses either as a hive or by direct instruction from a central program.

Nano-climites: These are specific devices which can be deployed to convert CO_2 into O_2 to supplement and regulate the base atmosphere.

Micro-machines: These are varying in size from small domestic appliances to insect dimensions. As with the smaller devices these can be used autonomously or in a 'swarm' configuration. One practical application for micro-machines is to maintain and remove dust and debris from solar cells and domes.

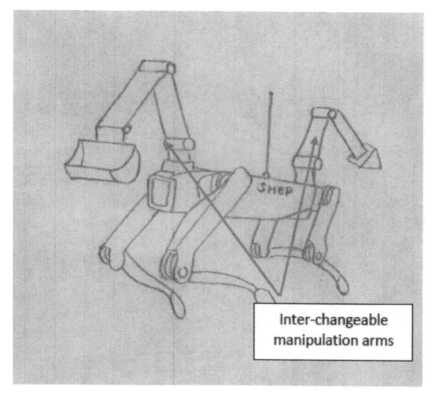

Figure 39. *Shep: Polypedal Robot*

Polypedal Robots: These will be deployed for tasks such as repair and maintenance but also for close confinement or exploration work. These include designs such as the 'Shep' robotic dog (see Figure 38).

This is a quadruped unit with added manipulation arms extending upwards allowing the fitting of various tool, sensors and equipment. The quadruped design provides a stable and secure platform and the arms allow for precise dexterity.

Cobots: Robots designed to assist in human tasks such as the bi-pedal Valkyrie R5 type design which has 3 fingers and an opposable thumb mechanism will be tasked to work alongside Martianauts and colonists. Designs such as the Robonaut 1 which is likened to a Centaur, featuring a wheeled body and human formed torso can also be tasked to work alongside human counterparts

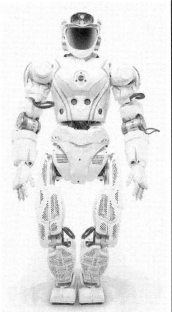 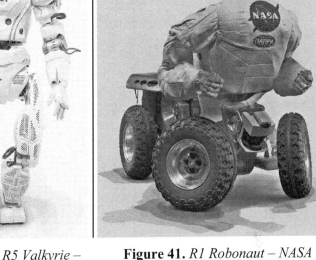

Figure 40. *R5 Valkyrie –*
NASA designed bi-pedal
robotic unit

Figure 41. *R1 Robonaut – NASA*
designed Centaur wheeled
robotic unit

9.2 Robot Applications

AI controlled machinery working to specific programming will be utilised for exploration and production tasks such as mineral mining and water extraction and processing as well as in the Hyperloop and SpaceElevator control platforms. There will also be the use of AGI (Artificial General Intelligence) and in addition to the use of Telepresence which will be employed on the surface.

9.2.1   Agriculture
- Monitor levels of
  - ○   radiation
  - ○ CO_2
  - ○   humidity
  - ○   crop growth and health
  - ○   soil nutrients, minerals and hydration levels

- Functions
  - rotavating and ploughing the soil
  - planting crops
  - hydrating the soil
  - providing additional minerals and nutrients
  - harvesting the crops
- Nano-climates (artificial plants) that will turn CO_2 into O_2 whenever it senses CO_2 levels are getting too high. These Nano-climates could be placed in every area where there is likely to be moderate to high levels of CO_2 such as bio-domes, work areas, vehicles etc.

9.2.2 Structural repair

- Gecko-like Mavroidisbots (using Van der Waal forces) will internally and externally monitor structural integrity. In the event of small punctures, proteins will be secreted to seal structures [21].
- Poly-pedal gecko-like will repair larger areas of the structure that the Mavroidisbots or humans can't fix or reach. All required tools will be integral.

9.2.3 Mining, Excavation and Dust Collection

- Excavators
- Ice borers
- Exploration
  - Every Explorerbot will have highly advanced neural networks and AGI incorporating fuzzy logic, evolutionary algorithms, reinforcement learning, search engines, knowledge bases and expert systems and will use GPS to navigate.

9.2.4 Health

- Medical Micro-bots
  - Nanobots carrying out medical procedures internally
  - Directed by surgeons or working autonomously.

9.2.5 Transport

- VR technology, telepresence control and telehaptics will be used by the astronauts within the base to drive transport vehicles and rovers for various purposes.
- Transport vehicles and rovers will also have VR technology, telepresence and telehaptics so that those in it can monitor and take control of Explorerbots that have been sent out from it.

9.2.6 Other applications

- Companionbot
- Petbot
- Tourism
- Equipment Repair and Maintenance

9.2 7 Integral Systems and Structure for Robots and Androids

- Micro-robot Support Subsystems
- Multifunctional Structures
- On-board Data Handling Subsystems
- On-board Power Systems
- On-board Thermal Control
- Telecommunications
- Beamed Power Generation
- Construction

10 Exploration and Resource Exploitation

10.1 Exploration objectives

It is intended that Mars will be explored from both Polar Base Alpha and Equatorial Base Bravo.

10.2 Exploration from Polar Base Alpha

The Martian polar caps are regions of scientific interest, but present challenges in their exploration. In particular, as on Earth, core samples will be taken from the ice cap to determine the effects of climate change and determine whether life ever existed on the planet.

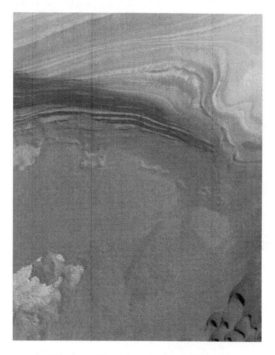

Figure 42. *Deep Spiral Chasm*

As noted by Cockell (2006) [29], there are two sites that would be of scientific and tourist interest:
- Chasma Boreale
- Spiral Valley

The Chasma Boreale is a visually stunning feature of the polar cap, consisting of a 520km valley with an average width of 60km and an average depth of 1.3km. Its origin is uncertain, with opinion split between Aeolian and fluvial. Either will be of scientific interest: If it is of aeolian origin, the explorers can examine those processes, whilst if it is fluvial, the origin of ancient meltwaters will be sought including determining whether they created habitable environments for life and the past Martian hydrologic cycle.

The Spiral Valleys are some of the most conspicuous features of the North Pole. They are approximately 1km deep, 10km wide and 300km long. The mean distance between them is some 50km. By cutting into the polar layered domains, they can be considered as a time capsule, offering access to samples from past epochs without the need for deep drilling. Expeditions will be expected to take 24 sols and will cover around 1,600km

10.3 Exploration from Equatorial Base Bravo

By being located at Meridiani Planum, Equatorial Base Bravo will be ideally situated for relatively easy access to numerous sites of scientific and tourist interest [30]. These will include:

- Firsoff Crater
- Sinus Meridiani
- Endeavour Crater

Figure 43. *Martian North Pole*

87

11 Emergency and Contingency Planning

As the colony will be as self-sufficient as possible there will be a need to ensure comprehensive facilities are available to meet the requirements of the population in the event of an accident, medical emergency or failure of the colony's life support systems. In line with this, suitably trained and equipped personnel will be required to be on station in order to deal with any such incident or emergency.

It is envisaged that all personnel will be fully trained in emergency first aid and that dedicated expert medical personnel will be stationed at Alpha, Bravo and SkyStation locations. These personnel will include:

- Medical technicians / Bio-technicians
- Advance Medics (AM)
- Advance Nurse Practitioners (ANP)
- Doctors (various disciplines)
- Psychologist

11.1 Pre-screening

To limit the risk of colonists arriving with existing medical conditions a rigorous pre-screening process will be carried out on all personnel. This screening and testing will include genetic marker analysis as well as mental and physical health checks. It will also record individual date line radiation protocols and establish individual health monitoring procedures depending on the individual and the task(s) for which they are being sent to the colony to carry out.

11.2 Emergency medical planning

As the transit time from the colony (SkyStation) to Earth can be up to 6 months as much medical treatment as possible needs to be carried out in situ. To this end two small emergency medical bays will be established on the SkyStation and at Equatorial Base Bravo with a fully equipped Medical Trauma Centre be situated at Polar Base Alpha.

The medical emergency bays will include the following facilities:

- Multi-use sterilizer equipment

- De-fibs / Resus equipment
- O_2 supplies
- Portable x-ray / Scanning equipment
- Ultrasound equipment
- Decontamination facilities
- Emergency trauma fixtures and fittings

In conjunction with this equipment the emergency medical bays will have appointed AM and ANP to staff them on a standby/ready basis. They will also run scheduled surgeries for minor consultations or referrals to the main Polar Base Alpha medical facilities. The AMs and ANPs will be assisted by robotic medical devices and through remote virtual assistance where required.

The Medical Trauma Centre (MTC) will be located at Polar Base Alpha and will be the main facility for the treatment and intensive treatment of colonists. To this end in addition to the facilities detailed above the MTC will contain:

Operation Theatre	Intensive Care Unit
Medical Lab	Isolation / Quarantine Unit
Robotic surgery equipment	Mental Health Facility
MRI Scanner	Resomator
Dental Suit	

When dealing with a casualty evacuation or rescue, the medical personnel will need to have access to emergency medical kit. These will fall into three main categories Basic, Advance Medic and Doctor. A general description of the medical kits is laid out in the following table with each kit containing the components of the preceding kit:

Basic Kit (all personnel)	Advance Medic	Doctor
Super Glue	•	•
Puncture Kit	•	•
1st Dressing	•	•
Straps	•	•
Auto-inject Morphine	•	•
O₂ Bottle / administrator	•	•
Plastic Cover	•	•
Spinal Board	•	•
2nd / 3rd Dressings		•
Endo-tubes		•
Chest drain		•
Lee Bore Needle		•
Scalpels		•
Hemestat Forceps		•
Air ways		•
O₂ cylinders / administration		•
Water Gel		•
Stretcher		•
Suction Pump		•
Fluids ᶜ/w admin set		•
Steri-strips		•
Suture kits		•
Meds & Creams		•
Hypothermia Kit		•
Radiation Kit		
Hypoxia Kit		
Major Surgery Kit		
Major Trauma Kit		

In the event of a medical emergency or evacuation the colony will operate the following protocol:

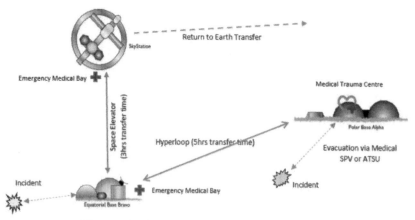

Figure 44. *Emergency Evacuation Protocol*

If an accident or incident occurs at a remote location, casualty evacuation and extraction will be via the use of a medical SPV. If the casualty requires stabilised transportation due to the nature of the injury and movement must be minimised, then utilisation of an ATSU will be employed (although slower). Any casualty transfers between Polar Base Alpha and Equatorial Base Bravo will ideally be done via the Hyperloop system as this is the fastest most direct route available. Transportation via SPV or ATSU between the bases is not a recommended option due to the distance (around 5 300km).

If the patient requires return to Earth transfer then they will be moved to the SkyStation for preparation. RTE transfer will then be requested or will be facilitated through the use of one of the 'lifeboats' attached to the SkyStation itself.

11.3 Monitoring and General Protocols

Along with the emergency medical procedures the colony contingency planning will also include more routine medical and environmental monitoring and testing to ensure emergencies are minimised as far as possible. Examples of these are:

- Routine testing of water system and hygiene systems
- Routine testing of food and food production processes
- Routine quality testing of air quality and LSS
- Monitoring of radiation doses
- Inoculations
- Medication and prescribing

Other non-emergency although serious conditions will have specific protocols as directed by the nature of the condition and will be dealt with primarily at the MTC and then if required via RTE transfer. Some of the conditions and protocols include:

- Pregnancy
- Radiation exposure
- Chronic debilitating conditions
- Debilitation resulting from injury
- Mental health issues
- Terminal conditions (DNR / P4 protocols)

Where there is a death of a colonist the protocol for the establishment of the cause of death will be the responsibility of the MTC clinicians. After the cause has been established the disposal of the body will be via the Resomator process. The Resomator will use Potassium Hydroxide and water heated to 150°C for complete breakdown of organic matter which could then be recycled into the colony bio-systems after neutralisation.

12 Governance.

It is proposed that Mars will, in effect, be owned equally by corporations and governments who will provide the funding for its exploration and development.

12.1 Corporate structure

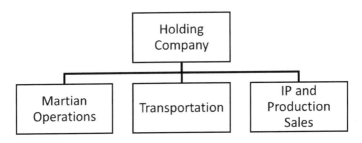

The **Holding Company** will be an Earth corporation. Its Chair and Board will be elected by the shareholders. It will be responsible for the overall strategic goals of the group, including ensuring that all required funding is available. Mars will be leased from the United Nations for 999 years by the Holding Company, to ensure benefit for all humanity. Shareholders are all of the companies and organisations that are involved with the Martian project, with each holding a single share and contributing operational funding on a proportionate basis based on the benefit they derive, after deduction of revenues received. Bonds will be issued to cover infrastructure capital investments.

In the medium to long term there should be excellent opportunities derived directly through profit from intellectual property (IP) patent licencing rights together with scope for ancillary benefits to be derived by participant companies from Martian operations that can be applied to other aspects of their businesses.

The Holding Company will be tax-exempt.

The model to be followed would be similar to that of colonial development charter corporations such as the East India Company and the British South Africa Company – albeit without the negative elements of those bodies.

The **Martian Operations** company will have a monopoly on all human operations on Mars. It will be a cost centre with costs accounted on a nett basis, with no profit element added. All financing required – both operating and investment – will be provided by the Holding Company.

The **Transportation** company will have a monopoly on transportation of personnel, tourists and goods between Earth and Mars. The company will operate combined passenger and cargo ships as well as freighters on a regular schedule between Earth's Lunar Base and the SkyStation. This will also be a cost centre with costs both accounted and charged out to users on a nett basis, with no profit element added. All financing required – both operating and investment – will be provided by the Holding Company.

The **IP and Production Sales** company will have the monopoly on sales of intellectual property rights for patents and systems developed from Martian operations, together with sales of minerals mined both on Mars and asteroids and manufactured products. All revenues from this operation will flow to the holding company.

12.2 Management Structure: Martian Operations

Mars will be a 'corporate planet' where all aspects of life are controlled by the Martian Operations company.

The corporate governance structure will be based on that of the German *Mittelstand* or medium sized business, where employee representatives make up at least half of the operating board, which is responsible for day-to-day tactical operations. Long term planning and strategic decisions are made by the Executive Board, which consists of shareholder nominees, non-executive experts and an employee delegate

The Chief Executive will have overall responsibility for the operations of the company and the welfare of everyone on Mars. This will include all disciplinary decisions.

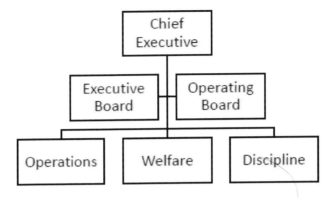

Employees and tourists will be able to purchase goods, including food, drink, souvenirs and luxury items through the company store.

12.3 Code of Conduct

The Chief Executive will fulfil the role of Executive President, and will be resident on Mars. The Executive Board will be based on Earth and will have oversight, while the Operating Board will be based on Mars and will have day-to-day executive responsibility.

Responsibility will be devolved for each of the key functions.

It is anticipated that for the safety and security of the entire planet and everyone on it, military disciple and command structure will be employed. In order to ensure that no individual or grouping has greater power or control than another, 'democratic' structures, although possible at a lower level, cannot exist at a supervisory one.

The Code of Conduct will provide comprehensive rules and regulations setting out both rights and responsibilities which will be binding upon both employees and visitors (including tourists) to Mars.

Appendix - Artwork

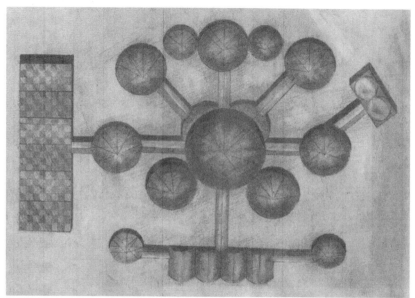

Figure 45 *Artist impression of Polar Base Alpha by Taylor D*

Figure 46. *Artist Impression of Polar Base Alpha's Reactor Unit and Greenhouses by Taylor D*

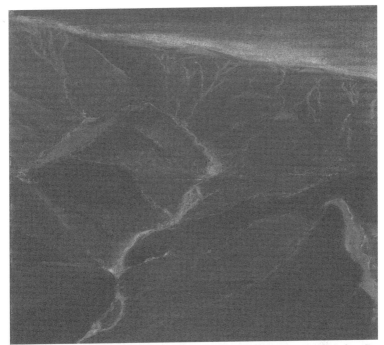

Figure 47. *Artist's Impression of Martian Valleys by Taylor D*

Figure 48. *Artist's Impression of Martian Landscape by Taylor D*

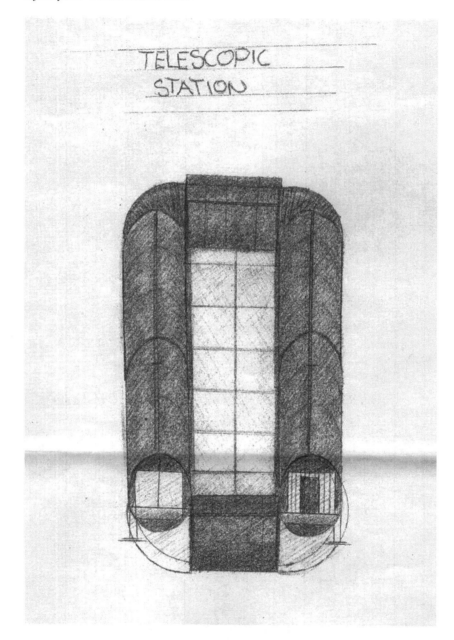

Figure 49. *Artist's Impression of Telescopic Habitat by Taylor D*

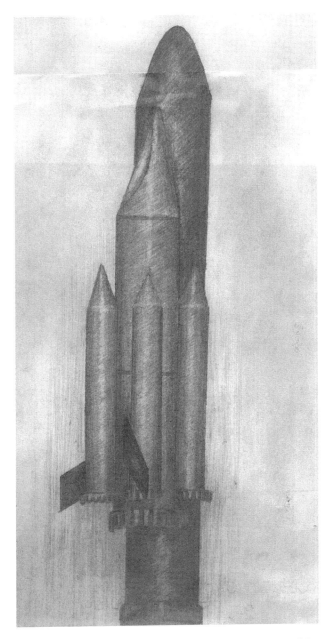

Figure 50. *Artist's Impression of Space Vehicle Launched from Bravo Launch Pad by Taylor D.*

Figure 51. *Terra Nova Logo by Taylor D*

Acknowledgements

We would like to thank Professor Cockell for giving us this unique opportunity to do something we never imagined we'd ever do in prison, and together with his PhD students Liam and James, for sharing their knowledge and enthusiasm with us. His help, support and encouragement has been invaluable. He has allowed us to show that people in prison have something to offer to the outside world.

We would also like to thank the SPS for allowing Professor Cockell and his students to do this project with us. Without their approval, none of this would have been possible. We hope that the SPS will continue to allow this type of project as well as any others that may arise in the future.

Appreciation must go to Fife College for giving us the space, time and resources for this project. We trust they will continue to support such projects in the future.

We thank our families and friends for their advice and assistance when we've spoken to them about it on visits, on the phone or in letters. They've shared their own interest in the subject and have helped to gather additional information.

We have all greatly enjoyed working on this project. It has led us to discover or rediscover interests and passions that may contribute to post-release employment prospects. While we recognise the limitations of life 'inside', we hope this project will encourage the ongoing provision of classes in the sciences.

Terra Nova had six contributors: John, Neil, Taylor, Martin, Gordon and John.

Unlike the majority of studies that look at a 'Day One' scenario for the initial human landing on Mars, we decided to take an innovative approach by looking fifty years beyond that point. All of the technology we have described is currently available, on the drawing board or will be available in the near-term.

John wrote the Life Beyond, Extra-planetary Infrastructure and Technical Solutions sections together with the Hyperloop subsection, including all of the illustrations. He also scanned and edited the drawings provided by other team members.

Neil was responsible for the compilation of this document and wrote the Human Factors, Habitat (together with John), Satellite Constellation, Transportation, Exploration & Resource Exploitation and Governance sections.

Taylor designed the Terra Nova logo together with the Mobile Telescopic Habitat and Polar Base Alpha designs and contributed Zulu chants to Ian's song *Martian Blues*.

Martin focused on robotics and wrote these Acknowledgements.

Gordon designed the Terrestrial Rover Vehicles, Mars Chopper Bike and Mars Railer.

John provided the Emergency and Contingency Planning section.

References

1. **D. Carrington.** Magic number for space pioneers calculated report on work of John Moore (University of Florida). *New Scientist.* http://archive.is/Xa8I.
2. **Peter Cattermole.** Introducing the planets and their moons. *Dunedin Academic Press.* 2014.
3. **Chris Impey.** Beyond – our future in space. *WW Norton Publishing.* 2016.
4. **Thorsten Denk.** SolarPACES conference. 2017.
5. **William Farrell.** Electric atmosphere on Martian moon. *Science Direct Journal.* NASA Goddard Space Flight Centre, So273117717305847.
6. **NASA Aerospace Scholars Program.** Mining and manufacturing on the moon. *NASA.* [Online] http://webarchive.org/web/20061206083416.
7. **P. Swan et al.** *An assessment of technological feasibility and the way forward.* 2013 : Houston Science Deck Books.
8. **Stephen J Gamble & The Space Exploration Team.** Project Boreas. [book auth.] C. S. Cockell. *Scheduling operations for the pole station.* 2006.
9. **C.S. Cockell, (Ed).** *Project Boreas: A Station for the Martian Geographic North Pole.* London : British Interplanetary Society, 2006.
10. **R. Zubrin.** *How to Live on Mars.* New York, NY : Three Rivers Press, 2008.
11. **John Badding.** Diamond Nanothreads. *Journal of Nature Materials.* 2015.
12. **Bryony Ashford & Dr. Xin Tu.** *Non-thermal plasma technology for the conversion of CO2* . s.l. : Department of Electrical Engineering & Electronics, University of Liverpool.
13. **J. Wang, G. Xia, A. Huang, S.L. Suib,.** CO_2 decomposition using glow discharge plasmas. *J. Catal.* 1999, Vol. 185.
14. **L. F. Gallimore and A. D. Spencer.** Efficiency of CO_2 dissociation in a radio frequency discharge. 2011.
15. **S. Paulussen, B. Verheyde et al.** Conversion of carbon dioxide to value added chemicals in atmospheric pressure dielectric barrier discharges. 2010.
16. **R. Aerts, W. Somers and A. Bogaerts.** *CO2 splitting in a dielectric barrier discharge plasma: A combined experimental and computational study.* 2014.
17. **M. H. Pham, V. Goujard, J. M. Tatibouet.** Activation of methane and carbon dioxide in a dielectric barrier discharge plasma reactor to produce hydrocarbons- influence of La2O3 / Al2O3 catalyst. 2011.
18. **James Rakocy, Michael Musser,.** Recirculating aquaculture tank production systems: Aquaponics – integrating fish and plant culture. [Online] 2013. www.aces.edu/dept/fisheries/aquaculture/documents/309884-SRAC454.pdf .
19. **University of Michigan.** *Space life and physical sciences: water to plant nutrient solution.* Ann Arbor : University of Michigan, 2017.
20. **M.A. Benjaminson, J.A. Gilcriest, M. Lorenz,.** Feasibility of an in vitro muscle protein production system (MPPS) for the fabrication of surrogate

muscle protein constructs as food products for space travellers. *Acta Astronautica.* 2002, Vol. 51, 12.

21. **Rod Liddle.** Finding an unlimited source of energy. *Sunday Times Magazine.* 19 November 2017.

22. **A Seltzman.** Design of an actively cooled grid system to improve efficiency in inertial electrostatic confinement fusion reactors. [Online] 2008. www.rtftechnologies.org/design/assets/device-images/fusormark3/files/seltzman_andrew_h_200805_phys.pdf.

23. **S. Atezeni, J. Meyer-ter-Vehn.** The Physics of Internal Fusion: Beam Plasma Interaction Hydrodynamics, Hot Dense Matter. [Online] 2004. http://fds.oup.com/pdf/0-19-856264-0.pdf.

24. **Ralph W. Moir.** Direct Energy Conversion in Fusion Reactors. [Online] 1997. http://askmar.com/fusion_files/directenergyconversioninfusionreactors.pdf.

25. **W.E. Horne, M.D. Morgan & S. Saban.** *Performance tuned radioisotope thermophotovoltaic space power system.* s.l. : NASA/JPL Caltech, 2008.

26. **X. Wang, et al,.** Prototype of radioisotope thermophotovoltaic system using photonic crystal spectral control. *Journal of Physics Conf. Ser.* Vol. 660, 012034.

27. **A. Schlock et al.** Design, Analysis and Optimisation of a RTPV Generator and its Applicability to an Illustrative Space Mission. *International Astronautical Fed Congress.* 2004.

28. **Martin Rees.** Where NASA Goes Next. *Wired UK.* November 2016, pp. 107-24.

29. **C.S. Cockell, (Ed).** *Martian Expedition Planning.* San Diego, CA : American Astronautical Society, 2004.

30. **B.G. Drake.** *Human Exploration of Mars: Design Reference Architecture 5.0.* Washington DC : NASA, 2009.

ELYSIUM STATION

Alan, Keith and David,
HMP Glenochil

INITIAL ELYSIUM STATION AND RATIONALE

Introduction

In October 2017, Professor Cockell from Edinburgh University came to Glenochil Prison to initiate a project on the theoretical design of a space station that could be transported and built on Mars. The prisoners were split into various working groups to look at possible designs. The parameters were quite wide, and given our lack of technical knowledge, we were to adopt a broad brush approach to the design. We were to look at basic design, drawings, paintings, stories, music, anything that could be included as part of a space project. Our group was a small group, consisting of three people. In the initial meeting, we set out what tasks each member of the group was to undertake and what research we were to do. In week two, we collectively started the design process and by week three, we had a design in place. In week four we were asked to look at how things would have moved on in 200 years from the time of the first Mars landing. The group chose to look at space tourism, and we chose to submit artwork of a space tourism centre on Mars, a poster to advertise space tourism and a list of all the major events that had happened on Mars in the 200 years since the first mission.

Overview - Why Go to Mars?

We identified several reasons to want to go to Mars:

1. To verify whether there is life on Mars or if not, whether there has ever been life on Mars.

The main reason to go to Mars initially would be to look for signs of life or whether life has ever existed on Mars. If life was found on Mars, you could then reasonably assume that if life was found on more than one planet within a solar system, then life would be commonplace within the universe. If life is not found, then life might be much less commonplace than initially hoped within the universe.

2, To search for rare minerals.

Once the infrastructure is in place, plant can be brought in to mine for rare minerals. However, we would foresee that to make such an operation viable, ore would have to be processed to turn the raw materials into the components required. This will be very expensive and we would envisage it would only be viable when all similar materials have been exhausted on the earth and moon. There may however be materials on Mars that are not available on earth.

3. To use Mars as a stepping stone for further exploration.

After the infrastructure is in place, Mars may be used as a stepping stone to explore the outer solar system and beyond.

4. To use Mars as a tourist destination.

Mars will ultimately become a tourist destination once the infrastructure is in place and travel times to the planet are reduced.

5. Because it's there.

Mars can be explored as an interesting location that is accessible.

Who should go to Mars Initially: Robots or Humans?

We considered whether Elysium Station should involve robots, humans or both. Given the distances and times involved and given that we are at the beginning of self- thinking and self-learning computers and that robotic technology is rapidly improving, it would appear sensible for the first mission to be totally robotic. This would be followed by a manned mission when the suitable infrastructure is set in place.

Advantages of Robots
We identified several advantages of robots:
1. Can survive and work in environments that humans cannot
2. No danger to humans
3. More cost efficient
4. Do not need down time
5. No accommodation needed
6. Easy to replace, maintain
7. Stronger, more hard wearing

Disadvantages of Robots
There are some disadvantages:
1. Robots do not have the ability to assess ongoing situation the same way humans can
2. Humans can evaluate differing situations and quickly solve multiple complex problems simultaneously, robots at present cannot, although we would see this imbalance changing in the future

Advantages of Humans
There are several advantages of humans that will make them useful in station construction:
1. To be able to evaluate differing situations and quickly solve multiple complex problems
2. To be able to assess the information collected and give opinions and options
3. To be able to use the biological information gathered by using humans for future deep space missions

Disadvantages of Humans

There are some disadvantages:
1. Danger to human life that a manned mission may pose
2. Humans have to carry their environment with them which makes a manned mission more dangerous and expensive
3. Not yet known what effect a lengthy mission will pose psychologically
4. Mission would have to factor in what effects prolonged exposure to radiation, less gravity, less sunlight and less human interaction would have on humans
5. Humans may need complex medical treatment which may be difficult to achieve on Mars
6. Humans need downtime, this will need to be factored in for the tasks involved

Should we go to Mars now or wait a few years?

We considered when the first Elysium Station should be built. Although technology is moving ahead at great speed, in our opinion we do not think we have the technology in place to go right now.
1. In a few years' time we will have self-learning, self-thinking robots to go to Mars first to set up the infrastructure that humans would need to survive
2. In the near future, plasma-powered rockets may become available which could greatly reduce the time taken to go to Mars

Given the distances and times involved, it would be practicable to wait until robot, rocket and computer technology improves. This would reduce the length of any Mars mission making it safer for humans. Improved robot and computer technology would ease the burden on what humans would have to undertake. It would also make the mission more cost effective.

Construction of Elysium Station: Initial Mars station missions required

We propose to send three missions to Mars, the first to be a totally robotic Pathfinder Mission. The second would be a manned mission containing five astronauts with engineering experience to build the station. The third mission would contain the scientists to run the station.

Pathfinder Mission

The Pathfinder Mission would be totally a robotic mission. Its purpose would be to reconnoitre the site where the station would be built. The purpose of this would be to survey the site, level the site if required, set out and electronically mark accurately the landing areas for the second mission. This mission would also contain robots, some materials and possible food required for the second mission.

Second Mission

Materials and personnel required

The basic requirements for the second mission are:

1. Five astronauts with engineering experience to supervise the landing of the landers and the building of the station
2. Four accommodation landing modules and one nuclear power generator landing module to ultimately power the station (Figure 1 and 2)
3. Robots to help build the station
4. Plants and seeds for future use when the station is built
5. Materials to build the station
6. Water purification treatment plants to take water from the permafrost
7. Air treatment plant to take oxygen from the Martian atmosphere and/or from the water stored in the ground
8. Sewage treatment plant

Objectives of Second Mission

This mission shall contain five people, and self-learning, self-thinking programmable robots to build the station.

Given the limit of weight of payloads that current rocket technology can deliver into space and given the weight and amount of machinery and materials required to build the station, it would appear sensible to launch several rockets with the required payload into Earth orbit and build the rocket to fly to Mars from the component parts. The advantage of this may be less fuel needed to get to Mars and human interaction on the long trip to Mars.

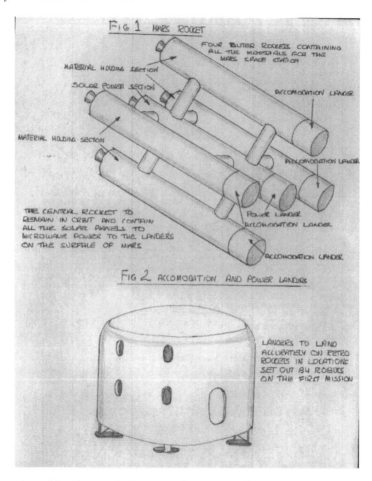

Figure 1 and 2. *Figure 1. Basic architecture of the spacecraft showing the four main landing accommodation units. Figure 2. Concept drawing of the landed station element.*

Once at Mars, the rocket would go into Mars orbit, where it would be de-assembled into its component parts. Four accommodation landers and the nuclear power lander would land on Mars using retro rockets on the sites accurately mapped out by the previous Pathfinder Mission.

The central rocket tip (Figure 4) containing the solar panels would remain in orbit and microwave power down to the landers until such time as the nuclear power plant is up and running, then the solar panels will be used as back up in the event of problems or maintenance to the nuclear power plant. The power would be used to supply the life support on the landers and the robots used to build the station.

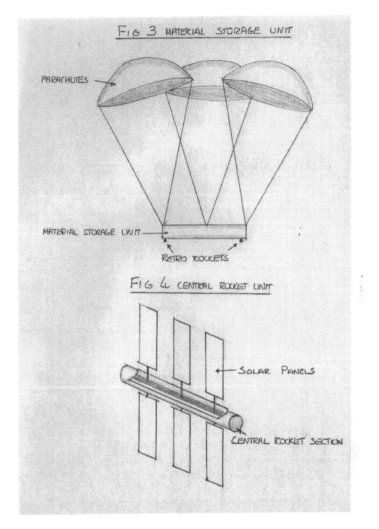

Figure 3 and 4. *Figure 3. Concept for material storage unit. Figure 4. Concept for central rocket unit*

The materials storage section of the rockets containing the materials to build the Mars station is to be parachuted down to Mars using positioning retro rockets to position the materials storage units within the vicinity of the site where the materials to build the Mars station can be unloaded (Figure 3).

111

Specifications of Elysium Station

The Mars space station shall consist of the accommodation unit (Figures 5 and 6), the science unit, the engineering workshop and medical centre housed around a circular walkway. The centre of the circular walkway is raised and covered to house the area to grow plants for food and oxygen. The original landers and the nuclear power plant are also connected to the circular walkway. Once built and sealed, oxygen is to be pumped throughout the whole unit.

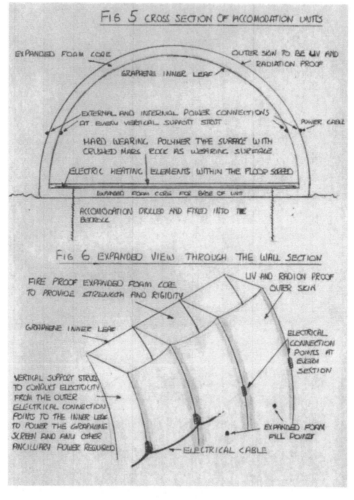

Figures 5 and 6. *Concept for the accommodation block and view of detailed structure.*

The units are to be lightweight, double skinned, units that are able to be collapsed and folded to save space to allow for more payload on the rockets.

1. The outer skin of the station to be a strong, webbed plastic which will be UV radiation and ionizing radiation proof with electrical connecting points at every cross-section.

2. The core of the units to be a carbon dioxide activated foam which is radiation proof, this will give the unit its strength and rigidity, and this will also act as an insulant.

3. The inner skin of the units is to be made of graphene, which will effectively provide all units with floor to ceiling computer screen, which can be used as lighting, wallpaper and computer screens at any point in the unit.

4. The cross-section supports shall conduct electricity and have electrical connectors in the external side and the graphene screen and electrical connectors in the inside thereby conducting electricity from outside to inside.

5. The unit is to be anchored through the floor into the bedrock.

6. The floor of the unit shall be a hard wearing polymer floor screed using Mars rock as the wearing part of its constituent make up.

7. Due to the low temperatures on Mars, the units shall require to be heated. The heating shall be a simple underfloor electrical heating cable contained within the polymer floor screed.

8. The graphene coated internal surface will be able to project any image or scenery, e.g. cameras on the external surface of the units could transmit images of the scenery outside to the graphene internal surface so the user would feel that they are walking outside. Alternatively, images of seashore or woodlands could be uploaded from Earth to give the people on Mars a taste of home. The graphene screen shall be used to provide all lighting.

9. Fire suppression would use Mars atmosphere to put out any flames. Fire doors throughout the unit will be airtight and remain in the open position to allow for air circulation. In the event of a fire within any section of the unit, the fire doors shall close automatically sealing the fire to that section allowing the fires suppression to take effect.

10. Oxygen is to be processed from the Mars atmosphere and or the permafrost (electrolysis), until the central unit is able to supply the oxygen to the station.

11. Water is to be taken from the permafrost and re-cycled and re-treated as much as possible.

12. All sewage is to be used as natural fertilizer. Waste water is to be treated and re-used.

13. As the strength of the sunlight is only about 40% of earth, electric lighting shall be required to assist the plant growth in the central growing area. Heating will also be require in the central growing area.

14. The central growing unit is to consist of a heavy see-through plastic type of cover, the cover shall be flexible to assist in its erection but the plastic shall become rigid upon activation by oxygen, giving the central unit a rigid shell, giving strength to the whole unit. Because this is see-through, this shall allow what sunlight there is to assist in the growing of plants to make oxygen. To make it more radiation proof, water may be run over the cover, this would be collected and re-used. The plants used for the production of oxygen and for food may be genetically modified to be less affected by radiation.

15. The unit at the centre of the growing unit shall be hydraulically operated allowing the unit to be built on the ground and then hydraulically lifted into the air (Figure 7). This would allow for easy access to all components, given that the engineers building the space station will have to be working in space suits, this will

greatly assist in the ease of build. The central unit shall house the air purification plant and air circulation unit.

16. An entertainment centre shall be provided within the accommodation unit. This shall consist of a room which would be able to project hologram images beamed from Earth to allow the user to have the illusion of going to their favourite football match or concert in real time or as near as the time delay will allow. This would improve morale and give the people in the station a connection with Earth. It could also be used to provide messages from friends and family back on Earth.

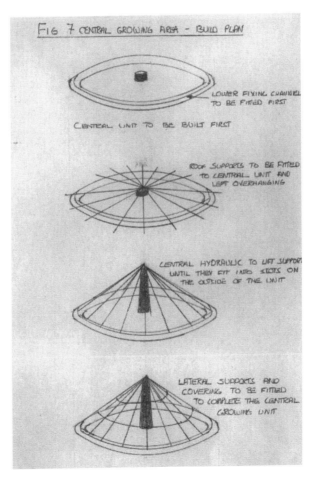

Figure 7. *Central dome section of station and concept of assembly.*

The Third Mission

The third mission would only take place once the space station was set up and in full working order. This final mission would contain all the personnel required to operate the station.

Figure 8-10 show the blueprints for the completed Elysium Station. Figure 11 shows an artistic rendering of the base as seen on Mars.

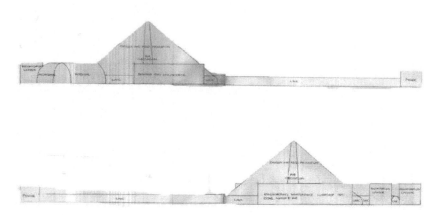

Figure 8. *Side-on view of Elysium Station*

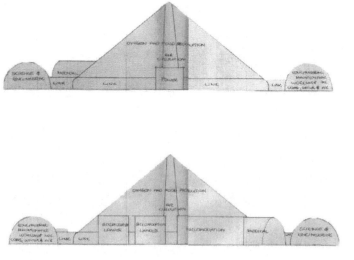

Figure 9. *Side-on view of Elysium main central hub.*

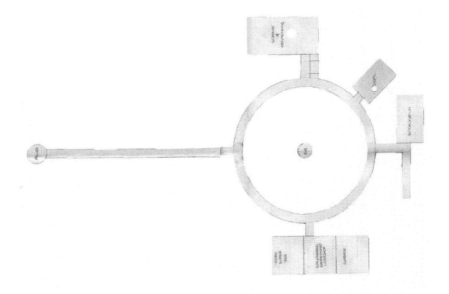

Figure 10. *Top-down view of Elysium station.*

Figure 11. *Artistic concept of the first Elysium Station*

MARS ELYSIUM HOLIDAY COMPLEX

The holiday complex is to be built entirely from materials sourced from Mars (Figure 1). It was built after the Mars oil refinery, plastics factory and steel mill was in production. It used steel for its main supports for both the outer shell and the core of the hotel units. Everything else is made from plastic in its various forms. The dome covering is made from an insulated and radiation-proof plastic developed specifically for the Mars environment.

Figure 1. *A painting of the Elysium holiday complex in the 23ʳᵈ century.*

Because all the materials are available on Mars, it makes the building of the holiday accommodation more cost effective. The majority of the construction was done by robots supervised by humans. Because of this, it was built to a high standard and in record time.

The power for the Elysium resort comes from a nuclear generator taken from Earth. The advantages of this system over solar panels is the guarantee of consistent energy not affected by dust storms or lack of sunlight.

The rich volcanic Martian soil is used for the golf course and all trees and plants, which are also used to augment the oxygen supply.

Most of the hotel staff at the resort are robotic with humans only in a supervising and engineering role.

The shuttle-craft takes tourists to and from their various destinations, taking them on excursions all over the planet (Figure 2), and to and from the Space Lift at the beginning and end of their holidays.

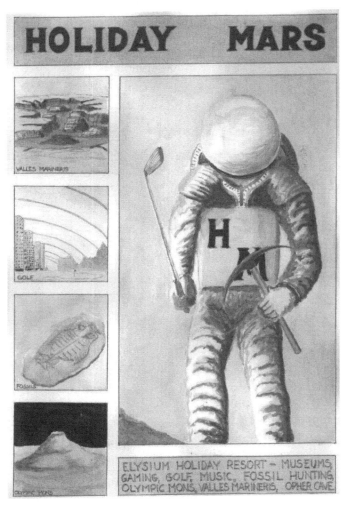

Figure 2. *A typical poster to be found at Elysium Station and on the Earth in the 23ʳᵈ century offering a variety of activities on Mars.*

ARS – A BRIEF HISTORY OF THE COLONISATION OF THE RED PLANET AND THE DEVELOPMENT OF ELYSIUM STATION

Glossary of terms

Elysium - The first settlement set up around the initial landing area.

Zakk Willis - The First human to set foot on Mars in 2030.

Deuterium - Heavy Isotope of Hydrogen (5 times more common on Mars than Earth)

Isidium - New type of mineral discovered in the Isidis region on Mars in 2107.

Amarsco - A mining conglomerate set up in 2040 to mine the red planet – backed by American, Russian and Chinese money and expertise (latterly to be taken over almost completely by the Chinese)

NASAX - New Company created in 2030 by the merging of NASA and Space X.

Viking X - Interstellar Space vehicles developed NASAX – capable of Earth to Mars travel in 6-8 weeks.

Elysium X 1 - Ship that exploded in transmit to Mars in 2159.

W.M.E.P. - World Mars Exploration Programme, world council for Mars exploration set up in 2051 as part of the U.N. to help oversee Mars projects.

M.C.P. (Mars Citizen Programme) – an initiative set up by the W.M.E.P. to populate Mars.

M.S.T. (Mars Settlement Treaty) – an agreement between nations to help formalise processes and procedures and ease tensions between Scientific, Industrial and Commercial operations on Mars.

E.S.E.A. (European Space Exploration Agency) – an off shoot of the European Space Agency set up specifically for Mars exploration.

A.M.P. (Asian Mars Project) – The Agency set up by Asian countries including South Korea, Japan and China. Predominately funded and controlled by Chinese money and interests.

M.I.C.P. (Mars Industrial & Commercial Programme) – worldwide initiative set up to utilise the new opportunities available on Mars.

Mars Embargo - A seven year period (2159-2163) of tension and uncertainty after the crash of Elysium X 1 – where almost all travel between Earth and Mars was stopped and the concept of colonising the Red Planet almost fell apart completely.

U.M.E.I.M.P (Unified Mars Exploration & Industrial Management Programme) – a new 25 year plan overseen by the U.N. and agreed in 2163 after two years of fraught negotiations between countries, companies, space agencies and the scientific community, which covers all aspects of Mars activity whether scientific, industrial or commercial.

Willis Space Inc. - Space tourism company set up by Bradley Willis (grandson of Zakk Willis) in 2090 to promote Mars as place for visiting and also to protect legacy of his grandfather.

M.N.S.O. (Mars Next Stage Outreach) - A programme set up in 2097 offering students and post graduates opportunities on Mars which include new short term stays of only four weeks on the red planet.

F.R.E. (Future Resources for Earth) – an initiative set up in 2111 to look at how Mars can help with Earth's global production.

M.W.U. (Mars Workers Union) – formed by miners in 2214.

Mars and Elysium Station – A Brief 200 year history

2030 After launching from Earth on the 18th March aboard the NASA ship "Explorer X", the initial landing party arrives on Mars in The Elysium region on the 12th September. American born Commander Zakk Willis steps out onto the surface of Mars, the first human being to stand on the red planet. He is quickly followed by Russian science officer Yuri Belchkin, Chinese engineer Lao Ti Sang and British astronaut Emma Currie. Pictures are beamed back live to Earth (with a time delay) to a world holding its breath as the four pioneers take a massive leap for humankind. The four are only on Mars for 17 days on this occasion but all of them will return in the future.

2031 The crew arrive back on Earth on the 30th March – their yearlong mission hailed a massive success and all agencies involved turn their attention to the next stage of Mars exploration.

2034 Robotic Mission – The initial robotic mission to Mars was to reconnoitre the site of the Elysium station. The mission objective was to survey the site, level out uneven areas and set out accurate landing areas for the manned mission, electronically mark the sites for the landers to set down on. This mission also contained materials and extra rations for the manned mission to use.

2036.. The 2nd Mars Manned mission with five people arrives on Mars in March and has a surface time of nine months – their mission is to build the Elysium settlement. The robots, food and propellant systems left in place by the original landing party in 2034 are all in good condition and the mission gets off to a great start, and despite some bad weather at times during the nine months the mission completes all of its objectives. The Elysium base is complete and this means that phase three can begin.

2037 The 3rd Mars Manned Mission – This mission followed the first when the planets align. The mission contains 12 people i.e. scientists, geologists etc. – who investigate Mars' past and present. They arrive in two landers, one of which was used by the five people already there to fly back to Earth, The scientists were able to start their main objective – the search for life and in late 2038 came the news, after the discovery of some fossil-like microbial organisms in Elysium, that scientists were able to confirm that Mars did sustain life in the distant past and theoretically could harbour some similar life now, but most probably in the deep subsurface only.

2039 A larger ship from Earth called the MayflowerX arrives on Mars in early October carrying the 16 brave first colonists of Mars. Their mission - building more accommodation units and continue exploration of the red planet.

2040-2065 More settlements built on Mars, mostly for scientists and Amarsco employees and their families. Amarsco set about geologically mapping Mars and begin setting up the infrastructure to start mining. While geo-mapping Mars, hydro-carbon layers were discovered, confirming that carbon based life had once developed on Mars.

2042 Only a few months before the end of the three year mission programme, there is an explosion in science lab 3 at the Elysium site killing German Scientist Angela Brehme who was only 34. The mission is halted temporarily while the other crew try to determine the cause and the safety and integrity of the rest of the settlement. Fortunately, Lab 3 is separate from the main dome and has not caused any further structural damage. Miss Brehme's body will be buried on Mars as she had requested. The W.M.E.P. decides to carry on with the next step in the colonisation programme despite the tragedy.

2043 Seven of the colonists who arrived in 2039 return to Earth as planned, months before the next ship is due to arrive. They return to Earth with a mine of scientific and geological information that will excite and engage Earth's scientific community for years to come.

The 2nd ship the "Weyland Yutani" lands on September 15th and carries a crew of 25 who will work along with the 12 crew members from the 2039 mission to carry on the scientific research and expansion of the settlement. They bring with them the latest in robotic automation technology which will enable work on the settlement to continue on an almost constant basis. This new batch of arrivals also includes geologists and representatives from Amarsco who will determine if mining works can be viable on Mars.

2046 A new U.N. backed organisation called The World Mars Exploration Programme (W.M.E.P.) is set up. This is essentially seen as world council for Mars – countries on the council include USA, Canada, China, Japan, The UK, Russia, Germany, Brazil, India and Australia.

2047 Confirmation of vast amounts of Deuterium and other areas of mining interest in the nearby region of Hesperia leads to Amarsco in conjunction with NASAX to commission the first major mining operation to be set up on Mars.

2052 The first working mine on Mars is officially opened it's called Hesperia 1 with plans for a major settlement to be built in place around the mine to house the workers and in future their families – in the short term the 75 workers will stay in temporary accommodation blocks with basic facilities, which have been pre-fabricated and brought to Mars in sections by NASAX.

2055 The first settlement at Hesperia officially opens with permanent accommodation blocks and facilities in place for around 100 workers.

2060 Amarsco announces plans to set up further mines and settlement areas at sites in the Isidis and Amazonis regions and also to increase the operations at the Hesperia site. This causes tensions between the scientific and industrial communities.

2063 A new agreement called the Mars Settlement Treaty (M.S.T.) is signed off by the Earth nations to formalise relations and processes between the needs of the industrial and scientific communities, and also to protect natural resources and areas of scientific interest.

The year also marks the death of the first human on Mars, Commander Zakk Willis who was 73 and spent almost 12 years of his life on Mars, across 5 different missions. A statue of him was erected at Elysium in his honour.

2066 The new Martians start to arrive, the first spacecraft "Viking X" carries in 425 people all of which have been selected from the USA's NASAX programme. This includes the first 50 people form the controversial "lottery winners". There are 264 men and 161 women and it is the first mission billed as a "one way ticket mission" with all those arriving expected to stay on Mars forever. People can leave if they want to though and of those 425 people who came only 64 returned to Earth permanently.

2068 Amarsco opens their first iron ore mine, the ore is of very high quality and the yield is high enough to make shipping back to Earth viable.

2070 Forty years after the first human set foot on the red planet, a Mars Government is set up. This allows decisions to be made quicker and more easily rather than having to run every little thing through the organisations back on Earth. However, its authority is limited and serves as not much more than a local council with all major planning still having to come through Earth.

2073 After a three year experimental process the first "Martian baby" is born – both parents were selected from major competition on Earth, sent to Mars and the baby was conceived and born on Mars and lived the first 18 months of its life on the red planet before making the journey back to Earth. The baby's name and details weren't revealed until 2082 in order to protect its identity. Baby "Zara Saldana" went on to live a full and healthy life on Earth only returning to Mars once, for a short stay in the late 2090's.

2074..New spacecraft designed by NASAX and with help from the A.M.P. takes over from the Viking X ships which have been in service for over 40 years. The new ships have more than double the capacity of the Viking X ships in terms of storage/cargo capacity, but with less room for people. They have a capacity of around 250 on the "Solis Class" crafts and 400 on the "Eridiana Class" crafts. The new craft enable more scheduled trips between Mars and Earth and shortens the journey time for major spacecraft to around 18/19 days. It also signals a change in direction as ships now offer more capacity for the booming cargo trade between the planets. Amarsco agree a major contract with the operators to enable them to increase the mining output on Mars.

2078 Drilling for oil began, leading to the discovery of a huge oil field in Isidis Planitia. The oil is of very high quality and has compounds within its structure not found on Earth.

2079 Plans were made to set up an oil refinery on Mars to produce carbon-based fuels and plastics to be used on Mars.

2085 Refinery and plastic production factory opens, producing fuel and plastics for use on Mars. Because of this development, costs for accommodation units fall, making living on Mars much more cost effective.

2091 First steel mill set up on Mars utilising the rich supply of iron ore on Mars. Steel to be mainly used for the building of accommodation units.

2100 The W.M.E.P. announce new plans to dramatically increase the settlement of Mars by lifting restrictions of those considered for the settlement programmes. This is called the Mars Citizen Programme (M.C.P.) and is due in part to the continuing over-population problem on Earth, which now sits at over 9.2 billion. The programme intends to send an initial 35,000 people to the red planet as part of a settlement plan over the next 15 years. In order to make this viable new interstellar space crafts are built with each capable of carrying 500 people from Earth to Mars in less than six Earth weeks. The original settlement at Elysium is considered the best site for the new expansion due to its superior infrastructure and, natural resources and potential living area. It's estimated that around 10,000 people will move to this settlement, taking its population to just over 12,000 by 2109. However, diplomatic tensions soon arise after the USA objects to the quotas of how many people each country gets to send. This leads to the USA temporarily withdrawing support for the M.C.P. for nearly two years and also leads to China taking on the leading vote role in the W.M.E.P. for the first time.

2102 The USA re-joins the W.M.E.P. and agrees a quota of around 1,500 people to join the M.C.P. They announce plans for a national lottery style programme which will pick 400 of the 1,500 people to go. All people who wish to enter the lottery must pay 25,000 Earth dollars and also pass various mental and physical tests in order to gain a chance of going. This causes tensions with other Council members including China and Germany who have insisted that all those who go to Mars are specially selected based on their scientific, academic and physical capabilities. This leads to further delays in the programme which has stalled badly since being announced in 2100. China withdraws full support for the plans.

More automated factories open, producing metal and plastic goods to be used on Mars.

2103..Seventy-three years after the first manned spacecraft arrived on Mars, around 6,000 people have been to the red planet, with a permanent settlement of around 4,000.

2107..A major scientific discovery of a new mineral in the Isidis region is billed as the Mars gold rush, the mineral is named as "Isidium" after the region and is seen as a major new component in the burgeoning robotics, cybernetics and artificial intelligence industry on Earth. This later became known as the first Mars boom.

The small mining operation already in Isidis is expanded ten-fold with plans in place for an operation with over 275 workers expected to be in place by the end of the year.

2111 Work eventually begins on the first part of the M.C.P. project at Elysium. China have re-joined the initiative. However, Germany and the UK both pull out for this first part due to financial concerns amid the tensions in Europe that have been building due to the ongoing energy crisis.

2114 The first expansion zone at Elysium called Elysium 1 is completed. All materials for this zone were manufactured on Mars. The new zone can accommodate 1,500 people. The infrastructure is billed as second to none with the latest technologies available. The area is almost self-sustainable with pod zones capable of providing modified crops and feed for the livestock programme including chickens, pigs and the marine zones which have been operating in Elysium over the last 10 years. It is estimated that around 75-90% of the food needed for the new arrivals will be harvested within the Elysium environment – drastically cutting down on the need for supplies from Earth which stood at around 55% of all food and sustenance consumed on Mars.

2120 There was a major mining disaster at Hesperia 1 when an underground explosion and mine collapse killed 31 people. Amarsco said that all safety issues and regulations followed on Earth were carried out ten-fold on Mars and the accident could not have been prevented. However, workers told of working practices at the mine as being very tough and in parts unsafe.

2123 A memorial is opened at Hesperia to commemorate those who died in 2120 – it is rumoured that Amarsco paid each of the victims' of the disasters families over $12m after admitting to some safety breaches.

2139 The population of Elysium hits 12,000 only 4 years behind schedule, there have been 16 different missions bringing people to the planet over the last four years and almost all nations of Earth are represented from the Americas, through to Europe and Asia. After some initial teething problems, the settlement becomes a success and plans are put into place to start working on a further two zones at Elysium. They have to be built over the next 20 years with workers from zone 1 involved in the building of the new zones using materials manufactured on Mars rather than all of this coming in from Earth – a major change from Zone 1.

2140 Mars gets its first "tourist attraction" with the original Elysium landing site and scientific/living areas being converted into a modern interactive museum with a holographic version of Commander Zakk Willis chosen as the tour guide.

2146 Amarsco unveil plans for further expansion at the mines in Isidis and also a major new operation to begin in the Amazonis region near Olympus Mons after further discoveries of major amounts of a new compound, Isidium.

2147 The scientific community gains an injunction from the W.M.E.P. to stop the mining project near Olympus Mons due to it being an area of major scientific interest.

2158 Elysium Zone 2 is now open and starts to receive the first of its expected 4,000 people, most of the arrivals are from Europe and Asia and is almost entirely privately funded with the colonists themselves paying for the privilege of going to stay on Mars.

2159 During the Elysium zone 2 expansion disaster strikes as the space craft "Elysium XM 1" jointly developed by the European Space Exploration Agency (E.S.E.A.) and the Asian Mars Programme (A.M.P.) suffers a major on board explosion, integrity breach and is lost in space. All 524 people on board die and no wreckage or bodies are ever recovered. During the investigation into the disaster, no exact cause is ever determined although it is thought that design flaws may have contributed to the disaster. Both the A.M.P. and E.S.E.A. are criticised for rushing through the design and building of the ship and the construction of Elysium XM2 and 3 are postponed indefinitely. The W.M.E.P. decides that all future building projects must be given the go ahead by them before any construction begins. Major new legislations are brought in to try and unify all construction and scientific procedures for all Mars programmes and to try to avert any future disasters.

2159-2163 The new legislation causes major problems and diplomatic tensions between the countries and companies involved in the Mars industry. Amarsco, now backed mainly by China, are the major dissenting voice and pull out of talks in 2086, vowing to carry on with their own individual Mars Programme regardless of support from the W.M.E.P. This causes major tension between those working within the M.I.C.P. and leads to the postponement of all Mars launches/missions which means that between 2086+2090 there are only three missions to and from Mars and only to send supplies and essentials to the red planet. This period became known as the Mars Embargo. The fallout from this causes the price of Isidium, which is now a major component on Earth in the manufacturing and engineering world, to increase tenfold in price as Earth based supplies start to dwindle. The U.N., W.M.E.P. and M.I.C.P. representatives gathered for major new talks in late 2161 to try to resolve the issues. After rather fraught negotiations, they eventually agree to a new 25 year initiative called the Unified Mars Exploration and Industrial Management Programme (U.M.E.I.M.P.) This new initiative takes nearly two years to negotiate and aims to cover all aspects of all activities on Mars including scientific, industrialisation, development and colonisation. The new programme has a few difficult moments but leads to many new technologies and advancements and signals a point of clearer thinking where for the first time since the Mars journey began all sides seem to be heading in the same direction with one clear aim.

2163 In late June, Amarsco in conjunction with NASAX re-commence travel between Earth and Mars using the previously successful "Viking X" spacecraft, bringing much needed supplies and fresh workers to the mines at Isidius and fresh supplies of Isidium back to Earth. The company also resubmits plans to the W.M.E.P. for mining works at the Amazonis site near Olympus Mons. Agreement was reached with the scientific community who had previously blocked the plans. The mining operation was scaled back and operations changed to take into account the areas listed as being of scientific interest.

2164 The new U.M.E.I.M.P. agreement is finally signed off and brought into law. To help mark this new dawn, memorials are opened simultaneously on Earth at the W.M.E.P. centres in Tokyo and Amsterdam and also at Elysium "to commemorate those who died when the ship Elysium XM1 exploded in transit to Mars in 2159. Recently gathered satellite evidence suggests that an undetected rogue meteor may have been involved with the loss of the ship, although this is never confirmed and conspiracy theories continue to surround the disaster.

2165 Viking X produces a prototype laser powered rocket, laser farms built on Earth and Mars to power the rocket. Trials are a great success. The rockets can under the right conditions travel at the fifth of the speed of light, and even allowing for accelerating and braking times they cut down the travelling time from 6-8 weeks to 6 hours.

2168 Full sized laser powered rockets developed by Viking X, the Viking X4, fly from Earth orbit to Mars orbit. Shuttles take people from the surface of Mars and Earth to the Viking X4 rockets waiting in orbit, making the journey time from surface to surface to under 8 hours.

2172 More arrivals at Elysium 2 mean that the colony is now full with a population of around 16,000 people. Plans are approved to finish the construction of Zone 3 which was left half completed when work halted in 2159 due to the Mars embargo. The zone will be completed in mid 2175 and will be able to accommodate another 3,000 inhabitants.

2173 Two new habitation zones near the mines at Hesperia and Amazonis are announced with work expected to start in early 2174 and be completed in four stages by 2196. This will see a colony with capacity for 12,000 people between the two settlements.

2178 Earth Lift and Mars Lift are developed by Viking X. In this construction, cables are stretched between the surface of the planet and an orbiting space station. They power a lift to take people into space, making it very cost effective to take people and materials into space, making space travel for the masses viable.

2184 Bradley Willis, great grandson of Mars pioneer Zakk Willis sets up Willis Industries which aims to protect his great grandfather's legacy and develop and promote tourism to Mars.

2207 A new scientific research settlement is announced near to Olympus Mons with a permanent colony housing 250 people and enough accommodation for another 250 visitors as part of a new scientific programme called the Mars Next Stage Outreach (M.N.S.O.). This will see students and newly qualified scientists complete short term stays on Mars which will last for around four weeks. With travel at only around three weeks either way, this will mean the shortest turnaround times of anyone who has travelled to Mars so far. The aim of this programme is to develop closer ties between the scientific communities of Mars and Earth and to develop further Mars commercial initiatives including Tourism. This is in part funded by a new consortium including Amarsco, NASAX and Willis Industries.

2214 With production at the three mining sites Isidis, Hesperia and Amazonis being increased to meet demands back on Earth, Amarsco runs into problems with workers calling for a planet-wide strike due to what they see as unsafe working practices. The workers join together to form the Mars Workers Union (M.W.U.) which calls for a halt to production until agreements are in place for an increase in wages and better facilities for its members. Amarsco agrees to negotiations but after an impasse is reached it threatens to replace the workers with new recruits from Earth. The stalemate lasts for nearly eight months after W.M.E.P. rules that Amarsco cannot just replace the workforce. The mining community on Mars is nearly 6,000 people strong by this point with around 85%+ members of the M.W.U. Eventually Amarsco agree to most of the demands and working practices change for the better with work resuming in early 2104.

2200 Mars permanent population hits 25,000.

2222 The new mining settlement areas at Hesperia and Amazonis are completed (three years late) and there is now living space and recreational facilities for nearly 12,000 people between the two sites. This will allow workers to set up home and for the first time have the option of moving their families to Mars if they wish to do so. Uptake of this offer is slow to start off with but Amarsco and the W.M.E.P. offer many incentives in the years to come and both sites soon reach their capacity.

2226 Mars now has four towns - Elysium (original settlement), Hesperia (mining town), Isidis (mining town), and Amazonis (mining town) as well as two small scientific outpost communities at Eridania and Olympus Mons.

2228 With the continuing global warming and energy problems on Earth, the scientific community push for more heavy industry to taken of Earth and moved into space. The Moon bases set up between 2065+2100 have been expanded to allow some of these works to take place there and Mars has been told to become self-sufficient in all areas. Food production on Mars soon hits 100% with hopes to start sending surplus back to Earth within 5 years. The Martian growing domes with their mixture of Earth and Martian soils have proved to be very fertile and efficient over the years and produce crops of good quality and longevity. Nearly 50% all building materials are now being produced on the planet and this will increase to 100% within five years. The mining towns of Hesperia and Amazonis will have new production and processing facilities added to them to allow this to happen. This will take away the need for costly pre-fabricated materials being sent from Earth.

The W.M.E.P. announces new plans to increase Mars population to around 100,000 people within 15 years – this is part of new version of the U.M.E.I.M.P. that includes plans for three new colonies, a new scientific programme further mining proposals and for the first time a bill of rights/constitution for Mars.

2230 The W.M.E.P. announces a new initiative called "Future Resources for Earth" (F.R.E). This is aimed at producing more resources on Mars that can be used on Earth and urges the scientific community to come up with new ways of using Mars to solve some of the population, global warming and energy issues that are causing so many problems on Earth at this time. This leads to what we call the second boom of Mars as Earth technology and scientific companies flood to Mars to try and find new, inventive and commercially viable ways to exploit the red planet.

2233 Just over 200 years after landing on the planet scientists in the Olympus Mons scientific research centre stun both worlds by announcing that they have found native current life on the red planet. For many years it has been thought that there was only signs of ancient life and some small possibility of microbial life across the planet. However advances in technology has allowed for further exploration of the red planet's hardest to access areas. Some worm-like creatures have been discovered in the deep subsurface beneath Olympus Mons. It is thought that these creatures have thrived in this environment for millions of years. The W.M.E.P. initiates its "life on Mars protocols" and the Area at Olympus Mons is put into quarantine whilst further research is carried out to determine if the new discoveries pose any threat to human life. The quarantine last for two years and the military is sent in to ensure all protocols are implemented. This causes delays in all ongoing Mars initiatives and for the first time in the years since coming to Mars there is a real question mark over the long term plans for Mars.

MARS STATION – THE EDINBURGH CONCEPT

Ross, Russell, John and Scott,
HMP Edinburgh

Abstract

We present a series of concepts and considerations for a Mars station. The design is based on a modular design construction that could be implemented in any environment on Mars. We discuss the rationale for the base, its design, how it could be used for science and tourism, and some aspects of governance and society at the station.

Keywords: Mars, modular, station design, research, tourism

Introduction

It has always been human nature to go exploring their surroundings. This has always been a means of survival, whether looking for food, better settlement, or to avoid potential disaster. Having explored much of Earth, there is not much we now don't know about our blue planet.

For centuries we have been casting our eyes towards the stars, wondering: What is out there?, Where have we come from?, and perhaps most importantly, Are we alone? Other questions of course arise like: Is our planet unique? Are there similar planets, habitable planets, out there?

With thanks to National Space Agencies, since the 1960's we have visited the moon, built an international space station, built large telescopes that see further into space and sent satellites beyond our planet, all in the name of Discovery

We have since discovered hundreds of new planets, stars and galaxies. We are just starting to understand what is out there in space and how vast it really is. Expansion seems to be the most natural thing since the Big Bang. Even asteroids and meteorites have arrived here from this great expanse, some crashing here on Earth, some burning up in the atmosphere and some of which grace us in passing by, called comets. But those fragments which we have found, we have analysed, we have found elements common on Earth but more surprisingly, amino acids essential to the building blocks of life as we know it.

We as humans love to explore and discover new things because of our natural curiosity, like landing a craft on an asteroid flying through space at thousands of miles an hour, analysing it and discovering we can do complex things like this.

We will inevitably endeavour to expand and colonise just like we did with new countries. We will no doubt do this with new worlds. This not only helps new discoveries and new knowledge, but will help with the survival of the Human Race. As our planet is quite populated, we are now looking to the stars once again for expansion, knowledge and the thrill of the unknown.

This thrill could start with a planet near our own, the red planet we call Mars! Someday we will have a research base on Mars, and years later expeditions to Mars for the common person and then colonisation. This will be the next step in human exploration. We will create new technologies that will help take us further into the cosmos and who knows, if there is any life out there, we will have announced firmly **WE ARE HERE, WE ARE READY,** and **WE ARE WILLING TO GO THAT EXTRA LIGHTYEAR!**

In the future we will have Earthlings and Martians but ultimately we will still be the Human Race.

So in the following pages, we present a few ideas of what might be needed to start this epic journey, one that starts with what is needed to live millions of miles away after a seven month flight (with current propulsion engines) as when we get there, it will take around two years to get back. A long time for any individual in the team, but also a short time in the survival of the human race, and a short hop to one planet that could start the foundation for countless others

Just like our predecessors, their gift of invention and discovery was a gift to us, our gift to future generations will be a new world, a new frontier upon which they could build upon and expand.

In this document, we will highlight some ideas of what may be needed for a base on Mars. The idea of this is to enable some research by a select few individuals to see whether living on Mars will become viable and hopefully one day promote tourism and colonisation of the red planet.

For this to happen there are some key areas that need to be considered such as:

- Where to build the base
- Power Sources / Fuel
- Research / Expeditions
- Staff
- Specifications of base
- Medical
- Farming / Self Sufficiency
- Leisure / Tourism / Affordability
- Life Support Systems
- Radiation Protection / Solar Flares / Cosmic rays
- Terraforming the landscape
- Generating Fuel
- Navigation / GPS Waypoints
- Weather – Devil Storms, Planetary sized Sand Storms (lasts weeks / months)
- Governance / Society

These areas all need to be considered along with possible funding budgets, length of time for the establishment, the psychological and physical effects, and affordability for the common person to gain such an experience in the future. We briefly consider each of these and in the process, present our own station design.

Where to build the base

The habitat could be built in an open area, which could help maximise ease of access, transport and potential for expansion. However, this could leave the base/habitat vulnerable to dust storms and habitants vulnerable to the solar radiation. The station can be built on stilts and on wheels to raise the habitat off the ground making it possible to relocate (this could be a smaller research base) to find the best possible area for further settlement.

This movement could also be due to seasonal change where in the warmer climate exploration could allow more a more northerly expedition (researching the polar caps and gathering ice for water) and in the colder climate an expedition to the equator. Or in times of a dust storm, relocating to a nearby cliff face for added protection of the habitat could be possible.

If the habitat is built in the Gusev Crater this can provide some form of protection from dust storms and solar radiation by having an extra dome built above the base modules. However, there will also have to be some form of access point to enable transportation in and out of the base/habitat.

Another possible site for building a habitat could be in the lava tubes of Olympus Mons as this can provide a large amount of protection from both dust storms and solar radiation. Although this volcano is largely believed to be extinct, more exploration may be needed to be sure there is no threat of any reactivation. Within the outer limits of Olympus Mons, there may be lava tubes that may be large enough to use as access points for transportation both for materials and access for tourists arriving and leaving the site.

In deciding where to build the research base, one of the factors that will need to be determined is the type of research that will be taking place. Once this has been decided, the above locations may be chosen from for the best purpose, distance and protective factors needed to succeed. However, other possible locations may be scouted to suit the purpose of this expedition or future expeditions.

Also in the future of the habitation of the planet, further suitable sites could be developed for more permanent housing, tourism and business growth in forming a new society on a new planet, for the expansion and survival of the human race.

Standard Modules

Each module should be a hexagonal shape to maximise space usage within the interior and prefabricated to enable an easy setup upon first landing on Mars to establish a base for the first team of researchers /

technicians for possible colonisation efforts. Each module will be connected with a tunnel that connects to the central hub once each module is established. These tunnels could also have a failsafe seal at either end to ensure safe keeping in the event of an engineering fault until such times that it can be repaired.

Each module could be an estimated size of around 250 square feet. This would give an approximate size of around 10 feet per side and around 7 high with a total of 6 sides. These should be prefabricated for easy use and quick setup with minimal effort needed using resources already available (i.e. robots for assembly or staff already inhabiting the planet). However, this may need to be adjusted depending on the location of the habitat if it is located within the lava tubes in Olympus Mons prior to excavation if any is needed.

These modules being of a standard size should enable the inhabitants the opportunity to customise the type of purpose that the module can provide. However, the central hub should be larger to provide a safe haven in the event of any unforeseen circumstances elsewhere in the station. It should hold major support systems that can be independent from the rest of the habitat to ensure survival.

The Individual Hubs

For safety reasons, each hub should be separated from each other in the event of an unforeseen circumstance, but connected via a tunnel / corridor that will have some sort of door seals. These corridors could be of a solid material or a flexible / inflatable tube. The separate hubs that should be considered are:

- Central Hub
- Living Quarters
- Medical Bay
- Power Centre
- Recreation Centre
- Research and Bio Lab
- Service Dock

The hubs may start for a research base, but when time has allowed they can be used for expansion to accommodate civilians and the growth of the society.

Central Hub

The Central Hub should be larger to first accommodate most of the amenities needed to establish the first habitable area for a select number of technicians that will be needed to build and establish a base area for further expansion.

This could mean that each side may be as big as 20 feet in length and 7 feet high with 6 sides. This should give a total of around 500 square feet. Within this area there will be enough space for sleeping, eating, communications, medical if needed and some space for research and repairs. This should be the first module sent to Mars with some machinery to help aid and move the module/s into position along with limited staff and provisions for an initial experiment to see whether colonisation on Mars is viable.

Following this being successful, we can then send further modules to expand the base. These may include separate modules for Research (Bio Lab), Medical Bay (with a possible quarantine area), Living Quarters for more teams or tourists to cohabitate (for establishing a colony), a Service Dock to enable engineers to fix and create new machinery essential to the living environment on Mars (this should be larger to accommodate vehicles needed for tourism, research and exploration), a much needed Power Centre (this could be a back-up generator and/or storage batteries to gather energy from a Solar Panel Field).

Living Quarters

The living quarters once placed in position should be connected via a tunnel that can be secured at each end with a door and seal in the event of any damage to the module to enable any essential repairs and maintenance to be carried out. This module should be of a standard size with easy connectivity to the tunnels with standard fittings that are of the same type and size to allow universal adaptability to other similar modules for expansion. This module should provide enough space for a sleeping area, washroom, kitchen and seating for downtime.

Life Support Systems

Each module within the habitat should have a standalone backup life support system that can be activated in an emergency, but also connected to the main support system that is controlled and maintained by engineers.

There should also be portable life support systems developed for use in vehicles and exploration. These should be constructed of a lightweight but durable material and designed to be carried with great ease should the need arise.

Oxygen/ Splitting Water

To get plentiful water on Mars, there would need to be an excursion to the polar caps for extraction of ice. Having retrieved ice, some can be used for melting for water and depending on how deep the drill station can go, some can be retained for further study into the past habitability of Mars itself.

The process for which to obtain oxygen from water is called electrolysis. This can be achieved by using two wires, a battery, some saline solution and a container with a lid to trap the oxygen. This will break down the water into hydrogen and oxygen

Another process by which this can be achieved is using a water solution of hydrogen peroxide. This can be done using two test tubes, hydrogen peroxide solution (3 or 5%), manganese oxide or active charcoal. The formula for this would be $2H_2O_2 \rightarrow 2H_2O + O_2$.

The above example can also be used to generate an alternate fuel source but this process may not be as efficient in terms of input and output.

This is just one example of how this can be achieved and there may be many more.

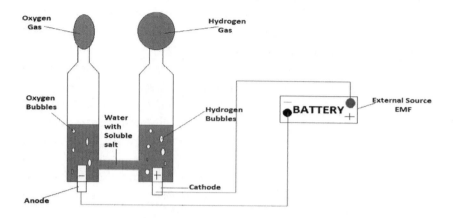

Figure 1. *A schematic example of electrolysis to be used for making oxygen in the station.*

Hydroponics and Farming

This area should be large enough to help produce a large yield of crops. This will help sustain a team of people long enough until further provisions are sent with the next mission. These crops can be genetically modified to grow quickly and produce a much larger yield. Eventually this will become sustainable with enough for farming and consumption. This will also help to create a sustainable and suitable environment for living and breathing. The hydroponics will help with irrigation, recycling and purifying water through filtering tanks.

These filtering tanks will also provide sanitation via chemical processes that will breakdown any biological waste for other purposes i.e. drinking water or fertiliser for crops via microbial bacteria.

The Solar Power Field and Power Centre

If the habitat is built in an open area, the Solar Panel field should be placed in an open area around 100 m from the base with easy to maintain panels (i.e. cleaning or replacement). This could be connected to the Power Centre through underground pipes to help protect cables from damage from dust storms. The panels could either be flat on the ground to avoid any debris that is blown in the sandstorms on Mars, but easy to clean with a brush or industrial street sweeper type machine / robot providing material used to make the panels are of a robust and strong nature (possibly using newly discovered Graphene). Or they could be built on rotating discs that could track the sun throughout the day to help maximise the amount of energy that can be captured.

If the base/habitat is built in the Gusev Crater with a dome above to protect it, then the solar panels could be placed on the dome to also help capture the sun light as it passes throughout the day, with an additional smaller solar panel field placed no more than 50 m from the closest access point to the site entrance, yet again either using a flat panel field or rotating discs to maximise solar energy capture.

If the habitat is built within the lava tubes of Olympus Mons then the Solar Power field will need to be built no more than 50 m from the closest access point in and out of the tubes. As Mars rotates there will be a shadow from Olympus Mons therefore it would be best to have rotating discs to track the sun throughout the day on two sides of the volcano.

Whichever of these locations that is used for the Solar power field, there should also be a connection to a storage facility that houses backup batteries to store the energy captured for use at times when solar activity is not viable.

Food- High protein, mineral and vitamin rich

The food supplies initially sent with the expedition to Mars should consist of high protein, mineral and vitamin-rich to ensure that the body gets all the nutrients needed. These supplies should also consist of seeds that can be planted and grown on Martian soil. These could be somewhat genetically modified for faster growth and a higher yield.

Terraforming

Terraforming Mars will take some considerable time on a large scale. However, for the Hydroponics and Farming area, this will be in a controlled area that will be covered with a see through dome to allow as much natural light as possible to reach the plants (to aid photosynthesis) along with lighting that will be hanging from the roof (heated bulbs, possibly UV). This will help with temperature control and growth for the crops.

Once the crops start growing, the process of photosynthesis (aided by irrigation) will help to provide an atmosphere within the building. After a substantial amount of time (many decades) and growth / expansion, this can be done at various locations with people to maintain and monitor plant growth that will eventually go towards producing an atmosphere of sorts with controlled environments. Following the success of this, the atmosphere will grow thicker with oxygen and perhaps in centuries to come, the planet will become as habitable as Earth without the aid of space suits or breathing apparatus.

For expansion of terraforming, we will need to use a plant that is the most efficient in converting CO_2 (carbon dioxide) and light energy into O_2 (oxygen) and likewise efficient in the use of minerals from the soil to aid growth. The most efficient way to send the trees to Mars would be by sending seeds although this will take time for the growth and nurturing of a small area.

Ice-producing water

With the ice that has been collected from the polar caps, this can be melted to produce water and then put through a purification process for human consumption. As not all the water produced will be for consumption, some will be used for sanitary purposes.

The water used for sanitary purposes can be recycled through filtration tanks using chemical and microbial organisms in the process to breakdown harmful bacteria thus reducing the amount of water wasted. Some of this water can also be used for irrigation in the farming hub or for terraforming with plants to help produce an atmosphere in selected areas.

The ice that is kept for research purposes can be used to study possible life or habitability on Mars, the history of Mars's weather systems and the effects it may have had on the environment.

Landing and Launch Site

This site should be accessible by vehicles traveling to and from a set distance of around 500 m – 1 km. This should be sufficient to enable safe launch and landing whilst also enabling quick transport and emergency rescue in the event of any unforeseen circumstances. This area could also have an outhouse type building with emergency supplies in the event of any unforeseen circumstances.

Medical Bay

This facility should provide all practical facilities needed in a medical emergency, such as an operating theatre, consultation area, examination room and an isolation chamber in the event of any unforeseen circumstances such as accidents, major trauma and unknown viruses. This facility should also be connected directly to a Research and Bio Lab in the event of such circumstances.

Service Dock

This service dock should be the first area in which any vehicle arrives and the last area that vehicles can leave. This dock or garage will have facilities to provide maintenance and fabrication of materials / equipment or engineering work that may need to be undertaken. This area could also provide temporary storage for resources gathered or received until needed. Within the service dock there can also be a portable drill station that can be deployed for research purposes.

Research and Bio Lab

This facility along with the medical bay will have an isolation chamber to enable safe research processes to take place whilst ensuring a safe environment for all habitants. This facility will include equipment for chemical and biological research along with personal computers for storage of results digitally and a cryogenic storage facility to preserve unknown viruses for further research when a time arises to enable scientists to do so. Another piece of equipment that should initially arrive with the first mission is a 3D printer to allow some replacement parts to be made cheaply and quickly in the event of a breakdown. In the event

145

of research that is far too dangerous, there should also be a chamber where the danger can be eliminated via cremation to destroy the threat thus allowing safe and continued habitation.

However in the event that this may not be possible there should be other fail safes in place to prevent contamination to the rest of the habitat such as being able to seal off the area via airlocks protecting the habitants.

This research can also be on the natural resources of the planet to help further the invention of new technology, generating fuel materials and sustainability of the habitat.

Recreation Centre

This unit should be around 30 feet in length and about 20 feet wide. This should provide plenty of space for some aerobics equipment and weights for training. In the centre there may also be enough space for a small court should habitants wish to play badminton or tennis (thus not restricting habitats to solo activities). This area could also be used for team building exercises to help build up skills and morale within the team. This module should be rectangular to maximise usage for sports availability.

Exploration / Navigation

In order to explore the planets surface, vehicles will need to be modified and fitted with stand-alone life support systems (or even a safety pod for larger vehicles), emergency space suits in case of any unforeseen circumstances, door seals that are airtight (possibly vacuum sealed) and a fuelling system that is economic (with a possible secondary power source) as any excursions may be far and wide from the base.

The same will be for navigation, as we have road maps and GPS for roads and countries on Earth, perhaps on Mars we may need waypoint finders / locators to help aid safe navigation on the Martian surface. These waypoint locators could also make it easier for future generations who inhabit Mars to find their way around and for any tourism agencies to promote interest. This will again aid the expansion and appeal to the human race on what could perhaps be a once in a lifetime event.

Generating Fuel

Having split water to obtain oxygen, a by-product that can be collected is hydrogen gas. This gas can then be used as an alternate fuel for combustion engines, cooking or possibly more. Please refer to figure 1 in the paragraph Oxygen / Splitting water.

There may be other gases from within Mars's atmosphere that may be able to be extracted for use in rocket fuel or for an alternate use.

This is just one example of how to generate alternate fuel sources, there are many more that can be considered.

Tourism

Figure 2. *Tourism poster prospect designed by Scott.*

As the affordability of space travel and living on Mars becomes cheaper and self-sustaining, the tourism industry will have a chance to grow. This will appeal further to the possibility of earthlings relocating to Mars in the future, thus promoting the expansion of the human race. Some of the possible sites for tourism may include:

- Gusev Crater
- Olympus Mons
- Arsia Mons
- Valles Marineris
- Hellas Basin
- Great Grand Canyon
- Giants Footprint
- Mars Pyramid
- Stone Circle
- Face on Mars

Following the success of tourism and expansion on Mars, there may be an opportunity to create and open theme parks that will further the interest to relocate to Mars. These theme parks could be unique to Mars due to the materials used through research and development, and the lack of gravity would produce a whole new experience.

Governance and Society

For the research base, governance could easily be settled by ranking, the same way as a military base is run for hierarchy. As this may be suitable for the initial base setup this may not suit a civilian setup when expansion has begun.

This is where a new system may be needed to be introduced for selection of individuals to provide oversight of new settlements in civilisation. How this may be achieved may depend on factors such as:

- Previous social status
- Qualifications
- Suitability
- Election from others
- Age and experience
- Previous occupations
- Length of term
- Laws to govern
- Accountability and Punishment

Then the election process - will it be by majority, selection or by one person proposing and one person seconding the motion? Will this process be democratic? Will this be decided before resettlement on Mars? This process will need to be set out clearly in order for the process to run smoothly and without any problems.

Summary of station design

Below is a proposed basic layout that can be adopted for a habitat in an open area or within the Gusev Crater. However, if the habitat is located within the lava tubes, this plan will need to be revised to best suit the size and location. The same with the location of the Power Field and possibly Hydroponics modules.

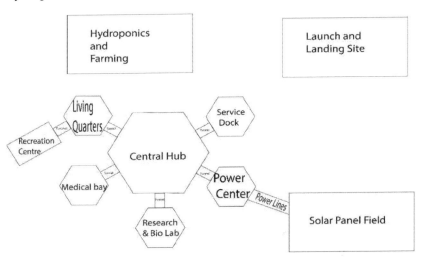

Figure 3. *Concept for the Mars Station. Please note that modules are not to scale with each other.*

When the habitat has been established, we could then setup a weather station to help predict seasons and sandstorms to help better understand the effects this will have on the environment in selected areas that are suitable for expansion. These weather stations should be able to monitor exposure to sunlight, solar radiation, sandstorms, wind and temperature.

In the research lab, experiments could be carried out on Martian soil to determine suitability for growth of crops and plantation. This in turn will help develop better irrigation, filtering systems using natural resources and creating an atmosphere within dome-like buildings.

These experiments will also help for expansion to suitable areas, which will enable tourism, affordability, cost effectiveness and meet funding needs for expansion. This can be achieved ideally after around 15 years of establishment of the first base on Mars. These first 15 years will provide time to carry out research for biological life forms / medicinal purposes, materials that can be used or discovered for use in new projects and give better insight into the planet's past for exploration and tourism in the coming years.

To populate this facility upon first being established there should be a select group of individuals (between 20 and 25 people), who will go through a screening process that will include tests for their expertise. These expert areas should include Medical, Physics, Engineering, Communication, Biology, Psychology, Chemistry and Geology. However, some of the applicants may have multiple areas of expertise which could help guide others if extra help is needed. This screening process should also highlight a hierarchy which will help maintain and govern the facilities in everyday life.

However, all of the above is a proposed research base design. This will need to be somewhat different for expansion when the time comes for the common person to visit or live on Mars in the future. For instance, the modules for housing may be of a slightly different design. Instead of having one floor there may be two floors like a standard house on Earth. They may be made of a newly discovered material on Mars which will make it cheaper to construct. Power sources may have advanced much further to utilise the planets geothermal sources, thus producing much cheaper energy. Vehicles will have been adapted for the planet's surface. These are just but a few of the ideas needed for expansion in the future and will make it more affordable and appealing to future generations.

Once the plans and designs for future expansion has been made and completed, new inhabitants start arriving on Mars. There will then need to be some viability to help grow and nurture this expansion, such as jobs that untrained personnel (trained after they arrive) could undertake to help maintain buildings, farms, vehicles, sanitation and help business flourish in the new settlements. No doubt after a few years (or even a decade, who knows) of settlers arriving, families will start to expand and then we will truly have the first real Martians within the human race.

These plans for the future expansion of any settlement should include the best possible location for equal opportunities of exploration of any possible site. This location could be near the equator to best suit any temperature or seasonal changes and distance travelled to any sites of interest.

Conclusion

There are many possibilities that could be considered if we had had more time to do so, with so many more ideas that could have been included in this document.

For instance, we could have considered how settlement design may be different, how the quality of life may be different than that on Earth, what possible effects will the lower gravity have on the day to day activities of life, and what new technologies or materials will we have by then and create whilst on Mars. Many more questions would still arise from the possible solutions to these questions.

However, this has been fun to consider in the meantime. With more allotted time, so much more could have been done to ask these questions and possibly answer them with even more possibilities for more creative ideas. There's no doubt in the future we will one day visit Mars in person and with new technological advances, we will travel further beyond.

Acknowledgements

This document has been inspired through a six week workshop in Astrobiology within HMP Edinburgh. Our group consisted of four members: Ross, Russell, John and Scott.

Within our group we were asked to come up with a base design for Mars that can possibly also be used for research, expansion and tourism in the near future of our species.

We came up with a few ideas and concepts, put them into a list and then split up the tasks to which we all took on. Ross took on the responsibility of compiling this document, expanding the ideas and drawing up a small but simple research base design

.

Scott took on the prospect of doing a speech that can be used in a presentation and a poster design for tourism.

Russell took on the task of creating an atmospheric theme tune called Martian Tribe using digital music software.

John contributed ideas and concepts to the initial design and theory of what may be needed to live on Mars.

MARS PROJECT CREATIVE WRITING COMPETITION

During the course, there were numerous discussions among the prisoners about what it would be like to live on Mars and to be the first people to start a life there. This led Alan in the Learning Centre to open up a creative writing competition to write the first email home from Mars. The exercise draws on several major questions – once you are on Mars, what things will become important? How will the experience shape your view of the planet and your future there? What would be the most likely things you would say to people back on Earth? The creative writing competition was an outstanding way to use the Mars project to contribute to literary skills and as the poster says, to 'let your imagination run wild.' The reader can see the impressive thought and imagination that went into the entries.

The entries begin with the winning entry and then are listed in no particular order.

You are working on the Mars space station and you are writing your first e-mail home. In no more than 500 words, what do you have to say?

In your piece, you can include anything Mars-themed that springs to mind. It's really up to you.

If you need a bit of inspiration, then you might want to consider your experiences of getting to Mars from Earth, about what it's been like living and working there, and perhaps about how you felt / now feel about remaining there. Let your imagination run wild.

Nanu! Nanu!

Deadline – Friday 8th December,

The winner will be announced during the week commencing Monday 18th December. There will be prizes for the top three.

Submissions to the Learning Centre Office 'for the attention of Alan'.

** NOTE ** - All entries will be forwarded to Professor Cockell and his team, to be considered for inclusion in a forthcoming book, showcasing the most innovative and stimulating creative and scientific work done as part of the HMP Glenochil / HMP Edinburgh Mars projects.

Figure 1. *The poster for HMP Glenochil advertising the **Mars Project Creative Writing Competition** organised by Alan MacFarlane in the Learning Centre.*

WINNING ENTRY

MAR(tyr)S

###TOP   SECRET###
NASA   Communications   Section
(message fragment #1068:69a)   {auto UHF packet: mars orbiter}
…
recipientidentified***disclosure***denied***referred***POTUS***

Maddie: this could be my last, you'll have seen the reports but it was no accident. This is coming by backchannel as no video/voice allowed/possible, the debris field has blocked all coms. You can't look outside, wreckage and desiccated body parts all around. The habi-module was taken out and we are holed up in the docking ring with juice for 20 hrs and tablets for the end. Everyone is too manic to care why a crazy man flipped out – or whether he survived it too! We are now 8.

But I had to tell someone it is so beautiful here, the reds are just so deep and rich, like spices; the shapes so familiar but alien for being untainted, fingers of waterless rivers stretch out across a new yet ancient world. Its depths are frozen iron, its air a poison, and its sky rains invisible death; but still we come with picks to rip up its skin and shovels to carry away its bones. They well name it the spoils, for nothing is beyond our greed and nothing will remain
.
They want people to sleep but they are scared the tablets are something else. The anxiety is eating the air and hastening the end, there isn't enough power to record messages so some are scribbling their last thoughts in the hope they will be found before our orbit decays – maybe decades from now. Cast adrift, a ship of death, testament to man's drive to reach out and crush all that is beautiful. Our footprint in the sands of a new world, the foreboding of a plague.

Doubtless we will be back, but now there is a window, to sue these 120 martyrs to prick the world's conscience, to solve our problems at home instead of dumping their decay on the virgin soil of new horizons. Use us as an outrage, use us as a threat, tear up our characters and expose our errors, tell them of our avarice; block, hinder, prevaricate, do what must be done. Keep the dream alive.

Tell the collective it was necessary, I had to sacrifice those people – shipmates and friends – but there was no other way. We are only 6 now and none suspect, it is better that way, I regret noth… (fragment ends).

*** recipient*** identified***disclosure***denied***referred***POTUS***

ENTRY TWO

To: Don.McLellan@Lunarops.nasa.org
From: Adam.Dimitriyev@Polarbase.mars
Date: 15 July 2045 02:45
Subj: We are not alone

Class: HIGHEST PRIORITY

This is the first email since I got here, and I guess it's going to be my last.

We've had manned bases on this planet for a decade and no one has found anything except sand, northern water ice, more sand, southern carbon dioxide ice and even more sand. The fricking stuff gets everywhere and into everything.   It zaps electrical systems, because it's magnetic.

Every winter there's humongous dust storms, resulting in total darkness that lasts for weeks. When Alex got lost last week, we thought he'd become disorientated and wandered off. But then the four guys in the search party also lost contact and didn't come back. We got a bit more worried, but not overly so. As I said, the stuff is magnetic and plays hell with navigation and comms systems, but the EVA suit life support system lasts for weeks. Remember when Bill Smith got lost in the Europe Mons cavern system – he was found three weeks later, physically fine but mentally...... well, I'm his replacement.

But this was different. John found the first suit near the nuclear plant this morning. He didn't recognise it at first – it had been completely abraded and mostly buried in sand. The helmet and other hard parts were scoured but otherwise undamaged. The weird thing was that there was no body, or even skeletal remains.

Jim and Kim found the second suit, also buried and worn almost to nothing. Again, no body. No sign of the other three. Francois checked the remains I the lab and that's when it got weirder. Nearly all the organic material had gone, leaving mostly synthetic fibres. Despite being in the microscope's vacuum chamber, when he switched off the light and went to infrared the individual grains of sand left on the suits began to move and clump together of their own volition. At first, the thought this was because they were magnetic, but then realised they moved to wherever cotton fibres remained. After a few moments, the clumps would roll on again and the fibres were gone.

It was consuming – *eating* – the fibres.

Next, he tested that with scraps of other organic materials – wood, cotton, rubber, wool. They were consumed too.

Finally, to check what had happened with the others, he tried a scrap of meat. The sand clumps went mad, attacking it like a shoal of piranhas, each grain dividing and multiplying. By the time it finished, the vacuum box was half full with sand – he says around a kilo of stuff.

The seals on all the airlock doors are rubber, and they're disappearing. Airleak alarms are going off, and there's more sand than normal inside here. As long as the lights are on, nothing happens. But when they go off, as they have been, because it's getting into the circuitry and fusing it, you can hear the stuff whistling and rustling as if it's been blown along – except there's no wind.

We're all together, waiting for whatever comes. So long Pal.

ENTRY THREE

Mars: 1st Email Home

Hello Earth, this is Mars calling, we've got our scores in and Jupiter nil point, only kidding haven't gone mad during travel. Talking about journey heRE, I recommend it to everyone, not only have I found a love for reading but I can now also speak Mandarin and God the sights we've seen. Tell the guy who ever coined the term empty space, well just him space is certainly not a cold dark empty place. Alright I'll give you that it's cold but so are the Alps and look how beautiful they are. I've downloaded pictures of our travels for you all to look at so I won't bore you like that mate that comes round with his holiday snaps, but wait till you check them out, I could say they are out of this world but I, oh I just have.

Technically there were no problems at toll on landing and setting up base, all went according to plan. Mentally, like I've mentioned with me reading a lot and learning to speak mandarin and also with all our tasks like the science experiments to do, keeping busy is definitely a big plus on the mental side. Physically, we all seemed to experience headaches about 6 months into the journey, meds helped initially but had little effect after 2-3 weeks, headaches not overly severe and could still carry out tasks and get sleep, all headaches have gone since arriving. Experienced joint pains like we thought we would once at base and in pressurised environment, but most gone and rest easing since we've started our pressurised workouts.

We hAve started to build Eden, made good progress and should be ready to plant first seeds on schedule. I can't wait for the first crop of spuds, not just for the project of work, though that is very important but, God I miss chips and mash and boiled and baked, must be the Irish in me. I really can't wait for Eden to be up and growing, how fantastic will that be? Growing life on Mars, it's a pity Bowie died before we got here or we could have told him, yes, David there is life on Mars. Though, if you did that, he'd have probably just told you to fuck off.

I've just been thinking that in another 100 years people might be looking back on this day as the first giant leap that enabled them to be on some planet another 1000,000,000 light years from Mars that they now call home. If that is the case, then let's hope that we have learned from our mistakes and that they take better care of their one than we have done with ours.

Well Earthlings, hope you like the pictures and I'll reconnect again soon, I'm just away to play the Soup Dragon at Ping Pong, so this your real life Martian saying, "to infinity and beyond".

ENTRY FOUR

<u>Mars to Earth</u>
Planchette-communique   (formerly   email)
Sent   6/1/3018

Hello Mary-Lou,

I travelled with nineteen others in the Exit Pod to the launch site, despite the training it did not prepare me for the launch into space, I felt I was being pushed backwards into my seat, good lord it was quite disturbing. Connecting with the space station we were transferred into the Travel-Pod which was luxury by comparison to the Exit Pod.

I never really understood what we were taught about the speeds we would reach with zero emissions in order to get to Mars, as we lay back I appeared to fall asleep as the air and temperature inside was reduced, when I awakened the date on my watch told me that only 5 days had passed, I felt a little time lagged however within 30 mins of touching down on the planet surface, we were picked up by an air conditioned bus which hovered along at quite a pace, entering into the station and we were shown to our room sand given a day to familiarise ourselves with our surroundings. The speeds we were travelling at were phenomenal leaving my skin feeling all prickly from head to foot.

I have now been here for a month and looking forward to the remaining five. It is fascinating to feel you are inside a goldfish bowl looking out on the planet, the surface is mainly flat, however, in the distance there are a few mountainous regions. My first day was spent at the health centre giving me a full health check, including a full Electrocardiogram (recording electric currents generated by my heartbeat), my vision, my hearing, my lungs including an Electroencephalogram (recording the electrical activities of my brain). I was then fitted with a microchip on the back of my hand under the skin which records my activities as I moved around the buildings, doubling up as a 'key' opening doors as the chip was scanned.

Some sights are truly transmogrifying when you first have vision of something unknown to you, realising that the unit generates its own purified air and water from recycled plant material. Seeing daylight for the first time is magnificent, it has a soft red hew on the horizon whether I was looking from within the station, or my 'outdoor' space suit, it was the same. I found this very confusing to my body, with little gravity feeling that I was about to 'take off' so to speak, feeling anxious I only remained outside in my suit for a mere five minutes, returning to the security of inside the station. I found it easier to breathe outside with less pressure on my lungs, less effort needed to move around the station than there was inside the manmade structure, being pure and controlled.

A huge part of the overall station, as much as one third of the coverage is committed to Xeriscapes and containing all kinds of varieties of Xerophytes. Quite strange to find when walking round that part of the complex, that there is no movement of air, suddenly I missed other things, there was no wind, nor even a slight breeze, I then missed the rustle of leaves, chirping of birds, flowing of water, all earth sounds were missing! Extraordinary.

Till next time X

ENTRY FIVE

To: LPC
From: TN5417

L,

This is my first and last permitted email.

It's been so long out here in the darkness.

I watched with awe and wonder at the clouds swirling around our iridescent cobalt blue home in the universe as we waited for departure, jostling with the other condemned at the tiny viewing portal looking down at all of you and wondering if you were looking up at me. Now home and you are a mere pale blue smidge of light with everything I have ever known or loved suspended shimmering within the hard vacuum of space, never to be seen again.

Mars, the underachieving ginger sibling of home now fills the view, rusty from decay, dusty from drought and awaiting its latest shipment of convicts. We huddle and jostle together again weightlessly nudging each other but now without awe or wonder just a morbid fascination of inevitability. The Elysium Planum is visible to us, smiling like some ironic joke; 'place of the blessed after death' is all there is and all there is going to be as we wait for transfer to the surface and from there to the mines, from where no return is possible
.

I wish I had done things differently, made different decisions and spared you from the anguish that has transpired. My happiest days and most treasured thoughts float around memories of you in the constellations and it is those which will comfort me in times ahead. You must now move on and embrace life in all its warmth and light and if you find yourself looking towards the horizon just after sunset remember the brightest star in the sky is where I am, sending my love back to you.

ENTRY SIX

MARS MY JOURNEY TO THE RED PLANET AND STARTING A NEW LIFE THERE

First of all Mars is miles away from earth. I was 6 years old when our family were picked to go. The scientist said it would usually take about 30 years to get there, but with this new jet system it halved the years. Then we were put into pots to hibernate, as I arrived at the planet with my parents we were given our quarters, my dad was a teacher and my mum was a nurse.

The year is 2050 we left earth on 2025 so that makes me 32, I became a scientist, things are amazing here my dad is 60 but he looks like 50, my mum is 55, she looks like 45. My job was to create synthetic food but everything I do is logged and stored in this massive computer.

I suppose living on the planet had its ups and downs, there were shops and bars you could walk to down the main street as I called it. The worst thing was you couldn't go outside the dome because there was no air, you would have to rent a suit if you wanted to go outside.

I have never been there because there are storms and they are very violent, anyway, I met this girl she was 28, we went out together for about a year, she was a nurse, the same as my mum, we decided to get married.

Name was Patricia and her dad was a police officer, her mum had died a long time ago, she said her mum couldn't cope with all the changes. She contracted cancer and with all the technology we have here they still can't find a cure for this terrible disease.
We went onto have two lovely kids, a boy and a girl. We called our son Leo and our daughter Libra, after the stars signs because my son was born in August and my daughter in November.

This is our life now every day the station is getting bigger. I got a new job it comes with a new quarters and it's taken me to the other side of the station. I also got a hydro car, my wife Pat wasn't very happy about leaving our house we had for 7 years, the only thing that is missing is nothing, as everything you could want is at your fingertips or at the press of a button or a phone call.

My wife got a job as a Tudor teaching nursing to students and giving lectures two nights a week. She also got a car so everything is looking and doing fine. My two children are students, Leo is studying to be a fitness instructor and Libra is studying chemistry so at the moment life is good on Mars, or what I used to call the red planet, because the earth is no more, man has destroyed himself.

ENTRY SEVEN

My journey on the probe to Mars

Day 18: So……. Great news folks! Am finally here, on the red planet at last. I can't tell you how proud it's made me feel and how happy it is to be up here right now, I can't seem to believe that am actually here on Mars. It's like I am in a dream, I can't shake myself from. It's taken seventeen days to get me here through tough challenges, immediate and swift actions and quick problem solving along the way.

My journey has been far from easy and as I am on my own, it's far harder than usual, especially a long mission like mine. Anyway after leaving Cape Canaveral, Florida, on launch day, one of the first things I had to do was disconnect the rocket from the boosters and probe. I'd pressed the release sequencer soon after leaving Earth's atmosphere, but unfortunately it was faulty and had to think of a solution quickly. I had to rip out the curing system and fuse the two rods together with tape. It worked, as I hoped it did. Once I had made sure everything was in order and the navigation system was heading for Mars. I then got some dried space food and did my daily exercises. After that I went to bed to hyper sleep.

When I woke up 16 days later, the master alarm had sounded to my attention, at some point when I had been sleeping, space debris had struck the side of the probe damaging some external navigation dishes which also took off the antenna for reaching contact with mission control. I had to quickly come up with ideas that would get me safely to Mars orbit with what I had. "I have a plan". My plan would be to navigate using the little viewing screens all around the probe which acted like rear view mirrors on a car. I would then need to find a way to contact earth with whatever I can. Once I was on Mars. "Maybe there was still old parts from previous unmanned missions I could put to good use", I said to myself.

When I saw Mars from the windows, I seen where I needed to land. "This is the real challenge", I told myself. I then put in the landing coordinates and handed the probe to the automatic pilot system.

167

When we touched down, I was so relieved at the fact I was alive. I shouted in excitement and I felt a massive sense of achievement, not for me alone but for the rest of mankind.

On the surface, it first seemed eerie to me but I got on with it and entered the space station. "Home sweet home" I yelled, I then found my bed and slept. I woke up hours later and walked outside to make the atmosphere processor. I then realised that the probe had vanished and no sign of it anywhere. I questioned myself "was this the storms that might have arrived overnight?" or "could this be the first encounter with life I have on this planet?"

AFTERWORD

Mark Hempsell
President
The British Interplanetary Society

I hope that you found something of interest and value in reading this book, which the British Interplanetary Society is proud to have published.

The Society was first introduced to the *Life Beyond* project during a presentation by Charles Cockell at a symposium on "Mars in the Age of New Space Launchers". To say it was an unusual paper would be something of an understatement. As well as explaining the overall project, he showed some of the output, highlighting the interesting perspective that being in prison provides - given the parallels with the enclosed environment of a Martian base.

When it came to discussing publication of the symposium proceedings Prof Cockell suggested a separate book publication of the project's output would work better to disseminate the work than a journal paper and the Society readily agreed. This unusual book is the result. I say "unusual" but really it is unique, which means the rationale for publishing it may not be immediately obvious. However I think there are two key reasons that amply justify its place in the Society's publication portfolio.

The first reason is its value to the overall literature of Martian bases – arguably a field that Prof. Cockell has led in modern times. The various teams have undertaken a comprehensive assessment of the requirements of advanced Martian bases. It is very rare for a group with no specialist skills to conduct such a considered review of Martian exploration from outside the astronautical world. Of course, it is not to be expected that this will compete directly with the many funded studies from teams of professional engineers working in large Space companies like Lockheed Martin, Boeing or Airbus. However, this work does complement these professional studies by providing a unique and very different perspective. So while it may not be essential reading for those considering the exploration of Mars, it is worthwhile, as an additional source of interesting approaches and ideas.

The second reason for the publication of the project's results is to highlight the work of the Scottish Prison Service and the contribution Space projects can make to meet its objectives to unlock potential, inspire change and build individual strength. It has clearly worked in this case within the Scottish Prison system's learning and skills programme. However, I think there is also a lesson here, in that it shows the ability of serious engagement of the general public with mankind's potential in space as a means to widen perspectives and inspire change in Society as a whole.

I know from Prof. Cockell's presentation that these studies have led to further work that is on-going at the time of writing. Some of this follow on work is original experimentation on aspects like growing food on Mars. So hopefully we can expect further interesting developments from the Scottish prisons' *Life Beyond* project.

ABOUT THE BRITISH INTERPLANETARY SOCIETY

The British Interplanetary Society (BIS), is a charity registered in the United Kingdom which is dedicated to the exploration and advocation of spaceflight. It was founded in the city of Liverpool, England in 1933 and has recently celebrated its 80th anniversary.

The BIS is Britain's leading think tank on space development. It is the world's longest established organization devoted solely to supporting and promoting the exploration of space and astronautics. It is financially independent, has charitable status, and obtains its main income from a worldwide membership.

The BIS is devoted to initiating, promoting and disseminating new concepts and technical information about space flight and astronautics through meetings, symposia, publications, visits and exhibitions. The Society's headquarters are located in Central London, within a 5 minute walk from local Underground, Rail and Bus Stations – enabling easy access for visitors to the Society.

A Brief History of the British Interplanetary Society

The British Interplanetary Society was founded by a group of space flight enthusiasts who dreamed of using rocket propulsion to fly to the Moon and the planets. However the word "Interplanetary" in the Society's title was not intended to limit attention only to those nearby worlds that circle our Sun, but to cover space exploration activities in general – even in the most distant visions to conceive of travel across interstellar space.

In the years before World War II a technical core of BIS members made first plans for a rocket capable of landing men on the Moon and returning them to Earth. After the war members of the Society developed ideas for the exploration of space, the construction of Space Stations, the human exploration of the Moon, the development of probes to investigate other planets in our Solar System and the use of space telescopes to observe distant stars and galaxies. Arthur C. Clarke first suggested the concept of communications satellites in a private memorandum to Fellows of the Society prior to first publication in 1945.

In 1951 the BIS organised the world's first International Congress on "The Artificial Satellite", and became one of the founder members of the International Astronautical Federation (IAF).

Planetary studies began with papers examining the propulsion requirements to reach Mars and Venus, as well as projected instruments for scientific probes. Amongst many examples of the Society's advocacy, studies on comets and meteors helped further the Giotto mission to Halley's Comet and later the ESA Rosetta mission. A number of papers during the 1990s also looked at the problems of making other planets (principally Mars) more hospitable to human habitation through terraforming.

Studies and long term thinking on human spaceflight and launch vehicle development have always been a core area of activity amongst Society members, with visionary early work on the BIS Lunar Lander, the Orbital Launcher and various other space launchers and space stations. More recently, it is from the minds of BIS Members and Fellows that have sprung concepts such as HOTOL, Interim HOTOL, SKYLON leading in some cases to hardware development.

With the demise of the Apollo Program in 1972, the British Interplanetary Society continued to encourage exploration of ideas on the possibilities of a return to the Moon and its eventual colonization.

The Society's first definitive paper on the requirements for Interstellar Travel appeared in its Journal as long ago as 1952. In the late 1970s a group of members worked together on the design of a possible interstellar vehicle that was published in 1978 as the Daedalus Study. From this followed many papers concerned with communications with extraterrestrial intelligences (SETI) and associated studies.

More recently a group within the Society initiated a series of studies on the scientific and technical objectives of conducting a human exploration of the Martian North Polar Cap (Project Boreas), and another group has initiated a follow on study to the Daedalus Interstellar Starprobe, called Project Icarus.

Membership

Membership of the British Interplanetary Society is open to all and intended for those with a general interest in astronautics.

Find out more about the British Interplanetary Society on their web site: www.bis-space.com.

Printed in Great Britain
by Amazon